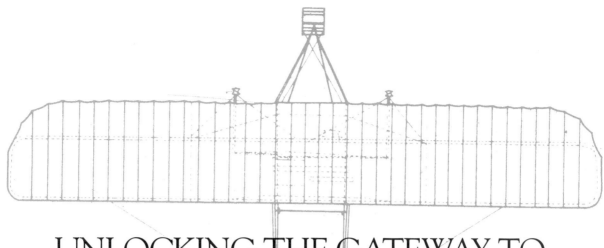

# UNLOCKING THE GATEWAY TO
# FLIGHT
## THE KEYS TO THE SUCCESS OF THE
# WRIGHT BROTHERS

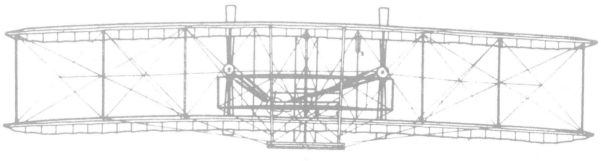

## BY

*Dale H. Whitford*

PUBLISHED BY THE WINKLER COMPANY
DAYTON, OHIO

*To*
*Sylvia*
*& our sons*
*Bob and Russell*

*With all my love*

# Table of Contents

This book is a publication of:
The Winkler Company
435 Patterson Road
Dayton, OH 45419

P.O. Box 572
Dayton, OH 45409
Telephone Orders: 937-294-2662
Fax Orders: 937-294-8375

www.Oakwoodregister.com

Manufactured in the United States of America by:
C&O Printing Inc.
Dayton, Ohio 45414

Library of Congress Cataloging-in-Publication Data

Whitford, Dale H. 1928 –
Unlocking the Gateway to Flight • The Keys to the Success of the Wright Brothers by Dale H. Whitford, with photographs and illustrations.

ISBN  0-9748415-052250(SB)
ISBN  0-9748415-153500(CB)

1. Wright Brothers • Wilbur Wright 1867–1912, Orville Wright 1871–1948
2. Invention of Flight
3. Aeronautical Engineering
4. Technology–History of Flight
5. Airplane

First Edition

# Acknowledgements

All aspects of writing and publishing this book have been enjoyable experiences because of the opportunity I had to meet and work with so many fine people. I want each of them or their heirs to realize how appreciative I am for their contributions.

Considerable first hand information about the Wright Brothers work and family life was obtained during interviews with their direct descendants: Ivonette Wright Miller, Horace Wright and his wife Susan, and with Wilkinson Wright. Their willingness to take the time to tell me about the Wright Brothers has been very helpful.

I am deeply indebted to my friend and mentor, Dr. Paul Garber, The first Curator of the Smithsonian Air and Space Museum, whose guidance and friendship was principally responsible for stimulating my interest in the Wright Brothers and for starting me on the lecture trail. Information in the form of drawings, slides, and photographs from him has provided the foundation for my research.

The kindness and encouragement of Amanda Wright Lane, Steven Wright, and Marion Wright, members of the Wright family currently living in the Dayton-Cincinnati area have been very much appreciated.

I am also indebted to Tom Crouch for the information I have obtained from his books and lectures. In addition, I am deeply grateful for the wealth of reference material I have obtained, particularly from the works of Marvin W. Mc Farland, Harry Combs, Peter L. Jakab, Howard Wolko, Rosamond Young, Howard DuFour, Fred C. Fisk and Marlin Todd. I also want to thank Fred Fisk for sharing information on the Wrights' bicycle business through the years at meetings and events where we have lectured together. Fred's suggestions concerning the publication of this book also has been helpful.

Ken Hyde, of The Wright Experience, has made an in-depth reverse engineering study of all facets of the Wright Brothers' work. His accurate reproduction of the 1903 Wright Flyer is scheduled to make the official flight at the First Flight Centennial at Kill Devil Hills, North Carolina, on December 17, 2003. I appreciate the information Ken has shared with me pertinent to the methods of construction and the efficiency of several Wright propellers. I also appreciate his willingness to share photographs taken at KittyHawk, North Carolina.

I want to thank Nancy Horlacher, Director of Curatorial Services at Carillon Historical Park, for her help in photographing The Wright Flyer III as well as the wind tunnel and other items in the Wright Bicycle Shop and the Object Theater.

The high-resolution photographs in this book were provided by Dawn Dewey and Toni Jeske, Special Collections and Archives of Wright State University; Betsey Knapp, Carillon Historical Park Archives and by Virginia and Robert Denison, who provided the excellent pictures they took at Kitty Hawk in November 2003. I am deeply appreciative of the contributions of these people and their prompt response to our needs

I am grateful to Dolores Winkler-Wagner, the publisher of the Oakwood Register newspaper, for publishing this book as well as assisting in the editing. In addition, I want to thank Lance Winkler, editor of the Oakwood Register, for his encouragement, support and editing of the manuscript.

I want the members of the publishing team to know that I really appreciate the product they have produced and that I have enjoyed working with them.

William J. Patterson was the leader of the publication effort and for all computer operations associated with distributing edited text, layout design, merging text and graphics and in generating prefatory material. He also acquired most of the high-resolution graphics. His expertise made it possible to produce the book in record time. I want to express my wholehearted appreciation for his contributions.

Bill Patterson, the author of numerous books requiring his expertise in Art and Photography, designed and composed all covers and produced all art and drawings and photographs in the from required for printing. I thank him very much for his outstanding contribution.

We all want to thank Larry McIntosh of C&O Printing, Inc. for his guidance during the early planning efforts in selecting quality materials and in coordinating and scheduling all functions of printing, binding and distribution.

And finally, I want to express my deep appreciation for the help and support by members of my family.

To our sons, Robert D. Whitford and Russell K. Whitford, I give my heartfelt thanks for writing various prefatory material, editing, and comments.

Above all, I am deeply indebted to my loving wife, Sylvia, who encouraged me to write the book, reviewed and edited the entire text and endured the turmoil and the "behind schedule" operation of the household in order to assure the book's timely completion.

Dale H. Whitford

November 2003

# Foreword

It was a cold morning on the beach. A Nor'easter had blown in during the night transforming rain puddles into blocks of ice and sweeping the land with strong gusty winds. The wind chill factor was nearly zero.

Despite the weather, two men slid a winged machine out of a shed and placed it on a level track facing the wind. After a brief conversation and a handshake, one of the men mounted the machine and turned on the engine.

Then at 10:35 in the morning, in the presence of five witnesses, the engine was revved to full speed, the machine rolled down the track, and in less than 60 feet it flew into the air!

The date was December 17, 1903. On that day, first, Orville Wright and then Wilbur Wright transformed the world by becoming the first human beings in history to successfully fly an airplane.

Few inventions have affected the world we live in as dramatically as the airplane. Yet the print media seems to delight in describing the airplane's humble inception in this trivial way:

*"The airplane was invented in Dayton, Ohio, by a couple of uneducated bicycle mechanics who tinkered around in their bicycle shop and somehow made an airplane that would fly"*.

No statement could be farther from the truth! These were well-educated, gifted men, who were two of the best engineers in the world at that time. Instead of "tinkering." they in fact conducted a classic research and development program that would rival the very best research conducted today.

They "made an airplane that would fly" not by luck, but by first reading and studying exhaustively to learn what was required for flight. Not the least of their problems was to survive the dangerous flight experiments so they could live long enough to learn how to fly. They solved the individual issues one by one until they achieved success; i.e. building and flying the world's first manned, powered, heavier-than-air, flying machine that was capable of sustained, controlled flight.

Neither of these two men had high school diplomas and both had only modest means. Yet, how did they, in just four short years, learn the secrets of flight that had evaded some of the world's most famous, well-educated, well-financed scientists and engineers for centuries?

This book was written partly to dispel some of the myths concerning the Wright Brothers and to describe in layman's terms the important technical and historical aspects of their work. But above all, it is the aim of this book to present those attributes, methods and achievements that constituted the Keys to the Success of the Wright Brothers. How they accomplished this feat of aeronautical engineering is an enthralling story about how the Wright Brothers became the first true aeronautical engineers.

November 2003

# Chapter 1 – Resources

The creation and development of any invention requires adequate resources to assure completion of the project. Wilbur and Orville Wright came from a midwestern family of modest means. Nevertheless, they were blessed with abundant and significant resources -- - solid family upbringing, good education, inherent skills and the necessary financial means required for successful fulfillment of their dream of achieving human flight in a powered flying machine.

## Key Number One •
## Family Resources.

The first key to the Wright's success was the enormous strength of knowledge and character they received from their parents as they grew to adulthood.

Wilbur Wright was born in Millville, Indiana on April 16, 1867. Millville is about 15 miles northwest of Richmond, Indiana on State Route 38. Orville Wright was a native Daytonian, born on August 19, 1871 in the family home at 7 Hawthorn Street.

Wilbur and Orville were sons of Milton Wright, and Susan Catherine Koerner Wright, Figure 1. Milton, a pastor and later a bishop in the United Brethren Church, was a loving yet strict parent who took an active part in rearing his family even though much of the time he was called to travel to the distant parishes which he served. He kept close contact with his family by mail and encouraged good behavior in his letters, just as he would have done if he were at home with them.

Although Milton Wright was a "preacher," he taught by both precept and example. He ingrained in his children a number of basic qualities, which were major keys to Wilbur and Orville's aviation success.

Milton taught them to self-educate by reading from his extensive library and other sources. He also taught them to keep journals of their work, as he did, and to earn their own spending money for all of their activities and business ventures.

But the most important lesson of all was that the Bishop taught his children to be independent thinkers, to have strength of character and to have confidence in their own abilities. Their father taught them that if they believed strongly in something and believed that it was right, they should not let anyone dissuade them, no matter how "high and mighty" that person might be.

Figure 1
From Left to Right: Wilbur, Katharine, Susan (mother), Lorin, Milton (father), Reuchlin, and Orville Wright

Susan Wright, their mother, also was an outstanding person. She attended Hartsville College where she excelled in mathematics, Greek and Latin. She also encouraged her children to read and pay attention to their schoolwork.

Susan's father was a carriage maker. Even as a child, woodworking and the use of tools fascinated her. She had great inventive and mechanical aptitudes, which she passed on to her children. She encouraged them to be inquisitive and to find out how mechanical devices worked by taking them apart.

Milton and Susan took advantage of every opportunity to improve their children's education. In their living room one can see in FIGURE 2 the door to a closet under the stairs. This closet was the detention center, where a child would have to serve time if he or she misbehaved. Interestingly, there was an outside window in the closet as well as a stack of magazines and books so a child could read even during detention.

When we talk about the Wright Brothers, we are inclined to think of two brothers. Actually, there were five sons and two daughters in the Wright family. In 1870 twins named Otis and Ida were born. Unfortunately Ida died at birth and Otis lived only a month. Thus, of the seven children, only four brothers and one sister lived to maturity.

Reuchlin, the oldest child, was born in 1861, and Lorin was born in 1862. Both married and had children. Wilbur and Orville never married and along with their sister Katharine, lived in their father's home, enabling the bishop to say he was honored that his three youngest children chose to "live beneath the paternal roof."

Katharine, the youngest child, was born on August 19, 1874, exactly three years to the day after Orville's birth. Katharine married late in life, well after the deaths of Wilbur and her father, but had no children.

I believe that Katharine was the heroine of the story. The Wright's mother, Susan, died when Katharine was only 15 years old. Thus, Katharine became the "lady of the house" and served the multiple roles of homemaker, sister, confidant and, at times, manager of the bicycle shop while Wilbur and Orville were in North Carolina conducting their flying experiments. She also was a major source of encouragement, urging them to continue their work when failures in their experiments tempted them to abandon their dream.

Many years after they had achieved aeronautical prominence, a friend of Orville's remarked to him:

*"Even though what you accomplished was without the idea of making money, the fact remains that the Wright brothers will always be favorite examples of how American lads with no special advantages can get ahead."*

To this Orville responded with much conviction:

*"But, that isn't true. Because, you see, we did have special advantages. We were lucky enough to grow up in a home environment where there was always much encouragement to children to pursue intellectual interests [and] to investigate whatever aroused curiosity. In a different kind of environment, our curiosity might have been nipped long before it could have borne fruit".*

In 1910, Wilbur had this to say:

*"If I were giving a young man advice as to how he might succeed in life, I would say to him: Pick out a good father and mother and begin life in Ohio"*

FIGURE 2
Door in Wright living room leads to closet detention center. There was a window in the closet and were plenty of magazines to read while in detention.

Figure 2

He followed his own advice!

Certainly, growing up in an environment headed by these strong, well-educated, encouraging and devoted parents provided the first Key to the Wright Brothers' success.

## Key Number Two •
## Educational Resources

There is a popular misconception that Wilbur and Orville Wright were uneducated bicycle mechanics. In reality, Wilbur and Orville had very good educational resources.

Wilbur was a quiet student. He did not volunteer much information unless asked directly by the teacher. Yet it became evident that he was very sure of himself. Once at age eleven, he had been severely scolded by a teacher because he had not obtained the correct answer to an arithmetic problem. So the teacher instructed a girl in the class to help him. They worked together for a while, only to discover that Wilbur's answer was correct. They returned to the teacher, and together convinced her that this was so. As time went on, it became evident that, indeed, Wilbur was a scholar.

As a senior at Richmond High School in Indiana, Wilbur undertook an exceptionally difficult program that would challenge gifted students of today. His curriculum included Greek, Latin, geometry, natural philosophy, geology, and composition. His scholastic averages were between 94 and 96 for the first three terms. His parents were very encouraged and started making plans for him to attend Yale to study for the ministry.

Unfortunately, Wilbur was not permitted to graduate from Richmond High School. His father made a decision to move to Dayton with so little notice that Wilbur could not attend either his final classes or his commencement exercises, though he had essentially completed his work. (Justice prevailed a few years ago, however, when Wilbur was posthumously awarded his Richmond High School diploma in Dayton).

Still with aspirations of going to Yale, Wilbur enrolled in Central High School in Dayton in order to enrich his education. There, he took additional courses in trigonometry, Cicero, and rhetoric. Central High School, which was located at Fourth and Wilkinson Streets, was eventually replaced by Steele High School.

One day, while playing ice hockey, Wilbur was hit in the mouth by a hockey stick. Several teeth were knocked out. After this accident, he suffered depression and heart palpitations. In his weakened condition, he was convinced that he might not live very long, and that it would be a waste of family resources to send him to Yale. So, giving up on a college education, he stayed at home, read profusely, and took care of his invalid mother, who was stricken by tuberculosis.

Wilbur was in a state of melancholy for about three years. During this time, he retreated into his father's library and diligently read books on history, ethics, science and the classics. He had an extraordinary memory, and thoroughly absorbed all that he read. After this three-year period, Wilbur had acquired an education that would rival anyone who had earned a Bachelor of Arts degree.

Orville's deportment in school was much different from that of Wilbur. He was very outgoing, mischievous and restless. Initially, Orville was a good student. In the first grade he won first prize in penmanship, and by memorization he advanced rapidly to third grade. However, during sixth grade in Richmond, Orville's mischievous behavior caused his teacher to send a note to his mother saying that he would not be allowed to return to school unless accompanied by a parent.

Orville was saved from punishment. His mother never had the chance to read the note because of the Bishop's abrupt decision to move to Dayton. Because he did not have a sixth grade certificate from his Richmond school, Orville was not promoted to the seventh grade in Dayton. Nevertheless he did well, earning the top sixth grade mathematics score in the entire city of Dayton.

By ninth grade, Orville's grades ranged generally from the high 70's to mid 80's, but in botany he had a grade of 92. He really liked botany, and enjoyed field trips that took him to Huffman Prairie where he studied rare wild flowers. (How important Huffman Prairie would be to him several years later!) Orville kept a notebook containing careful descriptions of plants, as well as beautifully detailed sketches of various flowers.

In tenth grade, Orville had raised his grades in mathematics, Latin and other subjects into the 90's. By that time it had became apparent that, while being a good student, his interest in entrepreneurial activities indicated he was destined to become a businessman.

Through the years, he had become interested in the printing business and constructed and purchased several printing presses. He even served as an apprentice in a printing shop during summers to prepare himself for the printing business. Before his senior year Orville dropped out of the regular high school curriculum and, instead, took advanced courses designed to prepare students for college. One of these was trigonometry, a course that would become very useful to him later. However, he did not have enough credits for graduation.

FIGURE 3 is a picture of Orville's senior class at Central High School, 1889-1890. Orville is at the center of the back row. Also in the back row on the left is Orville's friend, Paul Lawrence Dunbar who would become a world-famous black poet. Dunbar once wrote this poem about Orville:

*Orville Wright is out of sight*
*In the printing business.*
*No other mind is half as bright*
*As his'n is.*

From the standpoint of a formal education, neither Wilbur nor Orville received a high school diploma. However, they earned excellent grades in difficult curricula, took extra college preparatory courses, and Wilbur's three-year program of extensive reading was equivalent to a bachelor of arts degree. Having a good education was the second Key to the Wright Brothers' success.

## KEY NUMBER THREE •
## SKILL RESOURCES

The development of the first airplane required the coordinated, interdisciplinary abilities of a scientist, an engineer, a mechanic and an athlete.

The scientist studies and discovers the laws of nature. The engineer uses these laws of nature to design something that is useful. The mechanic builds the device that the engineer designs. And in the case of an airplane, the athlete has the physical coordination required to operate and control a machine, which is free to gyrate in three-dimensional space.

Most early aeronautical attempts were one-man efforts sometimes assisted only by a mechanic, or as in Sir George Cayley's case, his coachman. That poor man had never been given flying instructions prior to being ordered to fly. Consequently, most early "flying machines" failed partly because their developers lacked the flying skills necessary for success.

Fortunately, the Wright Brothers were able to approach airplane development in a completely competent manner. These were gifted men who, collectively, possessed all four of the necessary skills. Wilbur had a strong capability as a scholar and a scientist, and both brothers were excellent engineers.

Because of their bicycle business, they had gained knowledge of machines and mechanisms, as well as the machining, woodworking and welding techniques used in the manufacturing process. From use of their mother's sewing machine, they also learned how to fabricate wing and tail coverings.

Finally, both Wilbur and Orville were good athletes. Wilbur excelled as a gymnast

FIGURE 3
Orville Wright's Senior Class of 1889-1890, Central High School, Dayton

Figure 3

and football player in high school, and was reputed to be the swiftest runner in at Central High School. Both brothers were skilled bicycle riders. Orville even won several prizes competing in bicycle races. Their bicycle racing expertise transferred directly to their piloting ability; both men became excellent pilots.

These two men shared all four of these important skills, a condition that certainly promoted working efficiency. Today we accomplish this type of development by organizing a multidisciplinary team of people headed by a manager to ensure that all specialists work together in a coordinated way to meet final goals. Having all of the necessary skills was the third Key to the Wright Brothers' success.

## KEY NUMBER FOUR •
## FINANCIAL RESOURCES

The Wright Brothers were not wealthy men, but they were frugal and resourceful. All of their lives they had been trained to earn their own money for their activities and businesses. Thus, they learned to make use of discarded materials for their various projects. When Orville was 17, he, with the assistance of Wilbur, designed and manufactured a remarkable 1000-page per hour printing press made from a discarded folding buggy top, miscellaneous scrap parts and firewood.

The Wright Cycle Company, FIGURE 4, was thriving in 1899 when the brothers decided to study manned flight. And so, true to form, they planned to conduct their aeronautical experiments using only money they had earned from the surplus profits from their bicycle and printing businesses. Profits from both businesses provided not only for their own living expenses but sometimes for those of their father and sister as well.

Since they depended on business income for their livelihood, as well as their experiments, they were careful to avoid being carried away with their new interest to the detriment of their business and personal responsibilities. Hence, they initially conducted their experiments at Kitty Hawk in the fall and winter when inclement weather in Dayton caused a decline in customers' interest in bicycle riding.

With money limited, the Wrights were able to achieve their financial independence by contributing most of the required labor themselves, by being careful of expenditures and by strictly following a predetermined path toward their goal.

Financial independence was important to the brothers. Wealthy men had offered them financial support several times during their venture. This financing could have made it easier for them, and possibly they could have devoted more of their time during the bicycle season to their aeronautical activities. However, in all cases the brothers refused financial support. Why? By maintaining financial independence they also maintained complete project independence. No one had the right to tell them what to do, how to do it, or when to do it. And finally, when they were ultimately successful, they reaped all of the proceeds from their endeavors, that is, until they sold their patent rights and expertise to the American Wright Company in 1909.

Thus, the fourth Key to the Wright Brothers' success was that they had adequate and independent financial resources. In today's environment of government-sponsored research and development, it is amazing to us that the Wrights' made such a major technological advancement without external financial support, not even from the Government!

FIGURE 4
The Wright Cycle Company at 1127 W. Third Street, the source of funding for all of the Wrights' aeronautical research and development .

# Chapter 2 – Partnerships and Synergism

Being considerably younger than their other two brothers, Wilbur and Orville were drawn together very closely by the many joint activities they shared as they matured. The usual big-brother, little-brother relationship, arising from their four-year difference in age, began to change in childhood and continued to change as they became small businessmen involved in newspaper publishing, job-shop printing, bicycle manufacturing and repair, and finally the development of flying machines. From these ventures they developed the very important asset of synergism, which contributed greatly to their success.

## Youthful Ventures

As children Wilbur and Orville walked to school together. Wilbur taught Orville how to build kites that were better than any kites that could be purchased. Later, Wilbur helped Orville build a seven-foot, wood-turning lathe that worked very well.

While Orville was constantly looking for ways to make money, Wilbur had a great passion for reading. He absorbed all that he read and gradually showed a gift for writing.

Quite often his curiosity, as well as his interest in helping his younger brother, led him to participate in Orville's ventures.

When Orville was 12, he and two friends decided to stage a circus for the people in Richmond, Indiana. Wilbur was shocked when he learned, on the day before the big show, that Orville had not made plans to publicize the event. So, Wilbur offered his writing skills to compose a newspaper notice for the Richmond Evening Item. After studying various circus bills, he wrote a flamboyant notice with all of the extravagant verbiage used by circuses, describing both the show and the parade route that would be followed the next day. They anonymously submitted the notice to the newspaper office, and the editor's curiosity caused him to print it.

At parade time, the boys were astounded at the success of Wilbur's notice. The parade route was lined with so many people, the show had to be offered twice to meet demand. The show caused quite a stir among the adult townspeople;

*"--- the boy who had organized the show would doubtless amount to something"*,

some said. Others said that the youngster who had prepared the parade notice for the newspaper would surely be "heard from".

How "Wright" they were!

The brothers also were aware of their responsibility to the family. Together they added a spacious front porch to the house, turning all of the posts and spindles on their lathe, FIGURE 5. And together they remodeled the inside of their home, moving walls of various rooms.

## Printing Business

Although Orville started the Wright printing business with his friend Ed Sines, it wasn't long before Wilbur was participating. On two occasions Wilbur helped Orville build more advanced printing presses. After first becoming a casual writer and then the editor for the West Side News, FIGURE 6, Wilbur was invited to become a full partner in the printing business, Wright and Wright.

FIGURE 5
 Wright home on 7 Hawthorn Street, Dayton, Ohio

Figure 5 Courtesy of Special Collections and Archives, Wright State University

## Bicycle Business

The bicycle was very important in its day. The Wright's bicycle business certainly prepared them for their airplane development work.

Personal transportation in the 1890's was limited to walking, riding horse-drawn trams, in very limited cases, riding electric trolley cars, hiring a ride in a horse-drawn carriage or riding one's own horse or horse and buggy.

In 1878, the high-wheel ordinary bicycle was introduced. It had a huge front wheel and a very small rear wheel. The cycle was driven by pedal-cranks directly connected to the front wheel with no coaster mechanism. This bicycle was difficult to ride and its appeal was limited.

The safety bicycle, introduced from England in 1887, was similar to the bicycles we use today. It had two equal-sized wheels, and was driven by pedal-cranks mounted on the frame and linked to the rear wheel by a chain and sprocket system.

The safety bicycle immediately became the rage. For the first time people had means of personal transportation that was convenient to use, safe to ride, and financially feasible for most people. Men and women felt emancipated! They could take rides into the country, speed down a hill with hair blowing in the wind and would feel like they were flying. By 1895 bicycle production rose to more than one million per year. The "merry wheel" had become a national craze, and the bicycle was touted as being a "boon to all mankind," even a "national necessity." In fact, the bicycle had bridged the gap between the horse and buggy and the automobile.

By 1892, the craze had swept West Dayton and the Wright Brothers as well. For $160 Orville purchased a pneumatic-tired, Columbia bicycle. He believed himself to be a "scorcher" and loved bicycle racing, in which he won several medals. Wilbur, being more conservative, spent $80 for a used Eagle bike. He preferred rides in the country with Orville, where they met friends who also enjoyed the new pastime.

Figure 6

About this time, the printing business was declining. The competition they faced by the big city newspapers caused them to withdraw from the newspaper business, and the job shop printing business was neither interesting nor financially rewarding. Wilbur was bored, and Orville's interest floundered when compared to the exhilaration of bicycle racing.

Assessing their mechanical skills, the brothers decided to open a bicycle repair and sales business. They rented a store at 1005 W. Third Street in Dayton and stocked it with six brands of bicycles, plus spare parts. The Wright Cycle Exchange opened in the spring of 1893. Orville's friend, Ed Sines, helped keep the printing business going in conjunction with the bicycle business until 1898, FIGURE 7.

FIGURE 6
West Side News published by Orville Wright and edited by Wilbur Wright.

FIGURE 7
Ed Sines and Orville working in the bicycle shop

Figure 7 Courtesy of Special Collections and Archives, Wright State University

Business grew, requiring them to move from place to place, as more room was needed. By 1895, they had changed the business name to The Wright Cycle Company. According to Fred C. Fisk and Marlin W. Todd, the authors of *"The Wright Brothers From Bicycle To Biplane,"* the Wrights had six bicycle shop locations. It is also interesting to note that for a short time during 1895, one of the shops was located at 23 West Second Street in downtown Dayton next to the former location of Rike's Department store.

By 1896, while operating their business at 22 Williams Street, the Wrights realized they could make several bicycle design improvements. They also recognized the need to work during the winter when their cycling business was slow. So they decided to add a small machine shop and manufacture their own bicycle brands. True to form, they did just that and did all in their power to produce a high quality product that was reasonably priced. They named their original model the Van Cleve in honor of their early Dayton ancestors of whom their father was so proud. The Van Cleve originally sold for $60 to $65 dollars, and later for $55. Their second model, the St. Clair, sold for just $42.50. It was named for Benjamin St. Clair, the first governor of the Northwest Territories.

In 1897, the brothers moved their bicycle business for the last time into their shop at 1127 W. Third Street. Here, they converted the back room into a machine shop that contained an excellent lathe, a powered grinder and a drill press. They installed a line shaft from the ceiling with belts and pulleys that transferred the power from their one-cylinder engine to the machinery. They also obtained welding equipment for constructing bicycle frames. In building and using this workshop, they developed skills in woodworking, wood and metal machining, welding and designing and manufacturing mechanical equipment. This experience would prove invaluable to their aircraft business. From bicycle manufacturing, they learned about the use of ball bearings, gears, sprockets, chain drives and the fabrication of parts from seamless steel tubing.

The degree to which bicycle and bicycle-type components contributed to the building of the Wrights' gliders and airplanes is illustrated in the accompanying picture of the 1905 Wright Flyer, FIGURE 8.

Notice in FIGURE 8 that two large chains having one-inch-long links (much larger than bicycle chain) drove sprockets that were attached to the propellers. Sprockets attached to the engine drove the chain. Other chain and sprocket systems were used for camshaft drive, control wire pulleys and control operating mechanisms. Notice that the chain guides, the propeller shafts, the propeller shaft mounts and the radiator all required welded tubing construction. A large, flanged, bicycle-wheel hub was attached to the front of the plane's landing skid to keep the front of the plane rolling smoothly and securely on the launching rail during take off. Also there was a dolly made of ball-bearing wheels, which was placed underneath the plane at its balance point to support the plane and keep the rear part of the plane securely on the launching rail. And finally, one can see the welded tubing and the myriad of small welded steel fittings at the top and bottom of each vertical strut used to fasten the wings together and anchor the diagonal bracing wires. All of the skills required to manufacture and use these components were learned from the Wrights' bicycle shop experience.

Twenty-first century writers were not the first to see the direct link between the riding of a bicycle and the flying of an airplane. In The Bishop's Boys, Tom Crouch quotes James Means, a Bostonian who, in retirement, published a series of well-respected journals, *The Aeronautical Annual*. In 1896, Means noted the trend of people to equate cycling and flying. He said:

*"It is not uncommon for the cyclist, in the first flash of enthusiasm which quickly follows the unpleasantness of taming the steel steed, to remark: 'Wheeling is just like flying!'"*

Means urged those who sought to fly to pay serious attention to the bicycle. Once in the air, the operator of a flying machine would have to balance his craft and control its motion through the air. Balance, control and equilibrium were all problems

Two large chains having one-inch-long links (much larger than bicycle chain) drove sprockets that were attached to the propellers. Sprockets attached to the engine drove the chain. A chain and sprocket systems was used for camshaft drive.

Other chain and sprocket systems were used to control wire pulleys and control operating mechanisms. Notice that the chain guides, the propeller shafts, the propeller shaft mounts and the radiator all required welded tubing construction.

A large, flanged, bicycle-wheel hub was attached to the front of the plane's landing skid to keep the front of the plane rolling smoothly and securely on the launching rail during take off. One can see a myriad of small welded steel fittings at the top and bottom of each vertical strut used to fasten the wings together and anchor the diagonal bracing wires.

thoroughly familiar to the cyclist. Further, Means said that human beings would learn to fly just as they had learned to ride a bicycle: with practice.

"To learn to wheel one must learn to balance," Means pointed out. "To learn to fly one must learn to balance."

Within seven years, Wilbur and Orville Wright would prove the truth of those words.

## KEY NUMBER FIVE •
## SYNERGISM

The lifelong partnerships of the Wright brothers brought them so close to one another that they acted in many cases as one. The strong bond between them was best described in 1912 by Wilbur who wrote:

*"From the time we were little children, my brother Orville and myself lived together, played together, worked together, and, in fact thought together. We usually owned all of our toys in common, talked over our thoughts and aspirations so that nearly everything that was done in our lives has been the result of conversations, suggestions, and discussions between us."*

This statement leads us to their asset of synergism. Synergism occurs when two people work together in such a cooperative manner that the total effect of the two is greater than the sum of the two taken individually; i.e. two people working together so cooperatively that their work output is 1+1=3 or more.

We all have experienced synergism. When we build something requiring use of unwieldy materials, the work progresses much more rapidly and better when two people work cooperatively together.

Now consider the Wright brothers. They knew each other so well that each brother knew what the other was thinking, and each knew what the other brother was expecting of him. They had full confidence in one another, delegated authority well to each other, and worked together when joint work efforts were needed.

They both had an exact knowledge of the status of a project, what was needed and how the work of each would mesh constructively with that of the other. Thus their work equation might have been more like 1+1=5 or even 6.

Why was this trait important to their work? Their synergistic relationship enabled them to complete a given project in record time, eliminated waste and duplication, and reduce errors that would have been either costly, time consuming or even dangerous. Consequently, the fifth Key to the Wright Brothers' success was synergism.

The former President of Gates Lear Jet, Harry Combes, states in his book, Kill Devil Hill:

*"They worked so closely together, functioning as the superb intellectual team, that it never occurred to them to separate the financial matters of their business and their personal lives. As soon as they began producing income, they opened a joint bank account. Either could deposit or draw, which each did, without bothering to account to one another."*

Finally, in 1897, when it appeared that the "horseless carriage" would hurt the bicycle business, they looked for a new opportunity. The resulting joint venture became their last --- the aircraft business.

# Chapter 3 – Beginning of Flight Research

In 1878, Milton Wright was made a bishop in the United Brethren Church, making it necessary to move his family to Cedar Rapids, Iowa. Wilbur was 11 years old and Orville was seven. The Bishop had long recognized the value of educational toys for his children, so upon returning from one of his trips he had a surprise for his two youngest sons. As he approached Wilbur and Orville in the living room he held something clasped in his hands. He quickly opened his hands and a strange flying object fluttered out and climbed toward the ceiling. He had brought them a helicopter designed by the famous aviation experimentalist Alphonse Penaud, shown in Figure 9. It was a "bat", as they called it, which was made of paper and bamboo with a propeller powered by twisted rubber band. In its day, it was considered one of the best helicopter-type toys available.

The boys were delighted. After flying the toy and experimenting with it, they built several copies, which also flew. However, when they tried to make one twice as large, it wouldn't fly. They did not know why it failed and, like all toys that had passed their usefulness, set it aside and pursued other things. Orville in a court disposition later said that the gift of that toy was the true beginning of the brothers' interest in manned flight.

No doubt Wilbur and Orville read about all sorts of scientific achievements as they grew up, and we know that Orville, with Wilbur's assistance, was active in making kites. But from 1877 to 1892 the helicopter toy experience probably still glowed as an ember on the back burner in of their minds.

Otto Lilienthal, a German engineer, had been conducting aeronautical experiments for 12 years prior to 1891. During the next six years he began a comprehensive series of manned glider flight experiments, launching himself into the air from a 50-foot high, man-made hill as well as from the Rhinow Hills, about 50 miles from Berlin. The Wrights began reading newspaper accounts of his work beginning in 1892, and were mildly interested at first. Here was a man who was trying to do something that knowledgeable people said was impossible. Some scoffed at him. The glowing ember burned brighter, but still had not ignited.

By 1894, Wilbur was clearly becoming disinterested in the bicycle business. He wrote his father suggesting that he did not think he was fitted for a career in commerce, and that he should consider a teaching career, which meant he would need a college education. He felt he might be able to earn some of the required money from the bicycle shop. His father agreed with him and offered to help finance the college education. For some reason Wilbur did not follow through on the idea. Still, he was not content.

The spark of interest for flying was glowing brighter in Wilbur's mind. He spent much of his time thinking about the prospect of soaring through the air. He was particularly interested in spending time locally at a place called the Pinnacle Hills, in the City of Moraine, Ohio, just a little to the south of the Wrights' home in Dayton. Here were hills and sharp geological formations around which air currents undulated above the hilly landscape. Wilbur never tired of watching the vast numbers of birds that came there as they wheeled and soared effortlessly upward. Studying them through binoculars he watched and studied their gymnastics, always looking for the secrets of their aerobatics, which they performed so brilliantly.

## The Spark is Fanned

Then on August 9, 1896, shocking news came from Germany. After more than 2000 successful glides, Otto Lilienthal lost control of his glider while 50 feet in the air, and plunged to earth, breaking his spine. He died in a Berlin hospital the next day. His last words: --"sacrifices must be made."

There was much excitement in Dayton at that time, because the city was celebrating its centennial. Also near the end of August, Orville, became ill with typhoid fever, and nearly died. Consequently it wasn't until the end of August that Wilbur noted a short newspaper article saying that Lilienthal had

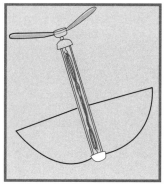

Figure 9

Figure 9
Toy helicopter designed by Alphonse Pènaud

died. To him this was a shock. More oxygen poured onto the ember. Wilbur realized that neither the world's best scientists in the past, nor even Lilienthal, had solved the problem of manned mechanical flight. Lilienthal's failure was particularly remarkable because he had studied the problem for twelve years, had made 2000 glides during a six-year period and was reputed to have collected the best set of aeronautical data in the world. Yet he had crashed to his death because he could not control his craft. This problem certainly would be a worthy challenge for Wilbur's mind. Perhaps this was the very endeavor for which he had been waiting.

Meanwhile, Orville's fever soared to 105.5 degrees; the family did all it could to help. Dr. Spitler, the family physician, said he had done all he could. There was no medicine. The fever would have to run its course.

Orville remained delirious and bedfast for six weeks until the fever finally broke. During this time, Wilbur spent time thinking about Lilienthal and the problem of flight. He even read books in the family library on bird flight. Finally, when Orville was able to sit up in bed and take light nourishment, Wilbur told him about Lilienthal.

The two brothers discussed the subject during Orville's convalescence, but when Orville became able they concentrated their efforts on the bicycle shop. Not much is known about their aeronautical pursuits in the two years following Lilienthal's death, but they probably continued reading and discussing the problem.

### The Spark Ignites

About 1911, Wilbur wrote the following brief in support of one of his patent infringement suits against the Glenn Curtiss Co.:

*"My brother and I became seriously interested in the problem of human flight in 1899.... We knew that men had by common consent adopted human flight as the standard of impossibility. When a man said -'it can't be done; a man might as well try to fly,'-he was understood as expressing the final limit of impossibility.*

Our own growing belief that man might nevertheless learn to fly was based on the idea that while thousands of the most dissimilar body structures, such as insects, fish, reptiles, birds, and mammals, were flying every day at pleasure, it was reasonable to suppose that man might also fly. We accordingly decided to write to the Smithsonian Institution and inquire for the best books relating to the subject.

## Key Number Six • Literature Search

With this decision, the brothers began their study in the classical way that mentors direct all researchers: i.e. collect, read and evaluate all that had been done before starting work on new research. Don't re-invent the wheel! This literature-search decision was the sixth key, which distinguished the Wrights' work from that of many other early aeronautical experimentalists.

On May 30, 1899, Wilbur wrote the following letter to the Smithsonian Institution, FIGURE 10

*"I have been interested in the problem of mechanical and human flight ever since as a boy I constructed a number of bats of various sizes after the style of Cayley's and Pènaud's machines. My observations since*

FIGURE 10
Wilbur Wright's letter to the Smithsonian Institution; May 30, 1899.

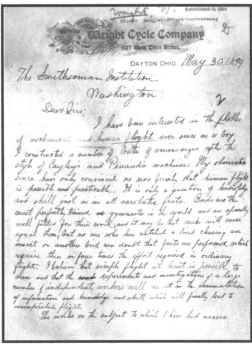

Figure 10 Courtesy of Carillon Historical Park Archives

*have only convinced me more firmly that human flight is possible and practicable. It is only a question of knowledge and skill just as in all acrobatic feats. Birds are the most perfectly trained gymnasts in the world and are specially well fitted for their work, and it may be that man will never equal them, but no one who has watched a bird chasing an insect or another bird can doubt that feats are performed which require three or four times the effort required in ordinary flight. I believe that simple flight at least is possible to man and that the experiments and investigations of a large number of independent workers will result in the accumulation of information and knowledge and skill which will finally lead to accomplished flight.*

*The works on the subject to which I have had access are Marey's and Jamieson's books published by Appleton's and various magazine and cyclopaedic articles. I am about to begin a systematic study of the subject in preparation for practical work to which I wish to obtain such papers as the Smithsonian Institution has published on this subject, and if possible a list of other works in print in the English language. I am an enthusiast, but not a crank in the sense that I have some pet theories as to the proper construction of a flying machine. I wish to avail myself of all that is already known and then if possible add my mite to help on the future worker who will attain final success. I do not know the terms on which you send out your publication but if you will inform me of the cost I will remit the price."*

Wilbur was thirty-two years old in the summer of 1899. He was convinced that the problem of manned flight offered him the opportunity to challenge his total capability. He was elated when, on June 2, 1899, he received a large package of information from the Smithsonian --- only three days after he had written to them.

In the package, Wilbur received reprints of articles that the Smithsonian had originally published. They were: Louis-Pierre Mouillard's *"Empire of the Air;"* Otto Lilienthal's *"The Problem of Flying and Practical Experiments in Soaring;"* Samuel P. Langley's *"The Story of Experiments in*

*Mechanical Flight;"* and E. C. Huffaker's *"On Soaring Flight."* In addition, the Smithsonian recommended that Wilbur acquire Octave Chanute's *"Progress in Flying Machines"*; Samuel P. Langley's *"Experiments in Aerodynamics"*; and *"The Aeronautical Annuals of 1895, 1896, and 1897"* by James Means. The receipt of these books set in motion a chain of events that would culminate in the invention that would transform the world.

At this time, Orville had only a mild interest in the flying problem. But as Wilbur continued to read and involve Orville into discussions of the work, Orville's interest began to grow until Wilbur had drawn Orville into the flying business just as Orville had drawn Wilbur into the printing business in 1889.

At about the time that Wilbur had received the Smithsonian package, Katherine came home from Oberlin College, bringing a female classmate. She had hoped that her brothers would help entertain her friend. To her dismay, they were so engrossed in their new reading material and related discussions, that they had little time for the girls. By beginning his research with a thorough study of what had gone before him, Wilbur was able to benefit from the lessons learned and the folly of others. Conducting a literature search at the very beginning of his project was the sixth Key to the Wright Brothers' success.

## Scientific Beginnings

Wilbur's reading essentially encompassed the history of aviation up to the year 1899, which is summarized in the following paragraphs.

In 1759, John Smeaton, a British engineer, was trying to learn how much pressure the wind exerts on the blades of a windmill. His measurements indicated that:

$$P = 0.005 V^2,$$

where:

**P = pressure in pounds per square foot**
and

$$V^2 = \text{airspeed squared in MPH squared.}$$

The term 0.005 became known as the Smeaton Coefficient in his honor. All 19th century experimenters also used this

FIGURE 11
Sir George Cayley's "Boy Carrier" glider

FIGURE 12
Alphonse Pènaud's rubber-band-powered model airplane

FIGURE 13
Samuel P. Langley's 16-foot, unmanned, steam-powered airplane launched from a house boat on the Potomac River

coefficient for computing lift and drag on airplane wings. The coefficient 0.005 was simply a measured value of the density of the air.

Sir George Cayley, later dubbed the Father of Aerial Navigation, is the man

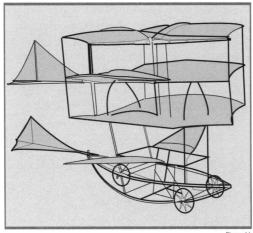

Figure 11

who really put heavier-than-air aeronautics on a firm scientific basis. He developed his principles using a whirling arm device to blow air over test wings so he could measure the lift and drag forces on them. He discovered that curved wing surfaces were superior to flat surfaces in creating lift. In 1804, he built and flew the world's first successful model glider, which looked very much like the gliders we build today, with a horizontal and vertical tail in the rear.

Figure 12

Cayley demonstrated his findings by flying a small boy in a glider in 1849, FIGURE 11. Later he coerced his coachman to board a glider and towed him into the air. After a somewhat hard landing the coachman declared he was giving his notice. He said he was "hired to drive - not to fly!"

Alphonse Pènaud is best known for his work in the design of aircraft having inherent stability; i.e. designing the wing and tail so that an airplane would automatically return to equilibrium when disturbed by a gust of wind. In 1871 he publicly validated his design by flying a rubber-band-powered model airplane a distance of 131 feet, FIGURE 12.

In 1866, Frances Wenham, a founding member of the Aeronautical Society of Great Britain, delivered a paper to the society describing his wind tunnel tests, which provided improved lift and drag data on several wing shapes. This work was very useful to the Wrights when they decided to conduct wind tunnel tests in 1901.

## Powered Machine Builders

Between 1887 and 1899 several men made full-sized, engine-powered airplanes. These included Felix du Temple (France), Alexander Mozhaisky (Russia), Clement Ader (France), and Sir Hiram Maxim (England). All of these planes had three things in common. They were steam-powered, they had no adequate control system and each succeeded in making only an uncontrolled hop off the ground. All except Ader's and Maxim's aircraft needed a downhill ramp to assist takeoff. Ader's first aircraft, the Éole, was said to have flown eight inches above the ground for 165 feet; yet his second and larger Avion III did

Figure 13

not even leave the ground. If one were to believe that any of these men could claim flight primacy, one would have to say that they were beaten by Eve in the Garden of Eden. Eve must have jumped a foot when she saw the snake!

Samuel Langley, secretary of the Smithsonian Institution had been conducting aeronautical experiments since 1870. He had studied bird feathers, bird flight and had conducted lift and drag tests of various objects. He also made several large, unsuccessful, steam-powered model airplanes. Finally, in 1896, his sixth model flew for three-quarters of a mile before it ran out of fuel, FIGURE 13. It had a wingspan of nearly 16 feet and was powered by a one-horsepower steam engine. By 1899, Langley had begun construction of a man-carrying plane funded by the U.S. government.

## The Flying Men

Of particular interest to Wilbur Wright was the work of Otto Lilienthal of Germany, Octave Chanute of United States, and Percy Pilcher of England. Of all these, Lilienthal's contributions were the greatest.

Lilienthal, an engineer, had been studying manned flight since 1879. He had made about 2000 glides in 16 different versions of gliders during a six-year period, beginning in 1891. Most of his gliders were monoplanes and some were biplanes. He flew either from a 50-foot high conical, earthen hill that he built near his home, or from nearby natural hills. Lilienthal's calculations and flight data comprised the largest set of flying-machine design information in the world. He also had developed formulas for computing lift and drag of wings. As mentioned before, his work was cut short on August 9, 1896, after he died in a glider crash.

Octave Chanute was a famous, wealthy civil engineer. In semi-retirement he pursued aeronautics as a serious hobby. He designed gliders, which were built and flown by his laboratory assistants from the dunes near his home in Chicago. He also was an information disseminator. He traveled throughout the world talking to leading aeronautical experimentalists and sharing information. He documented his findings in his book, *Progress Flying Machines*, which was considered the "bible" of aviation. Chanute promoted the use of a biplane wing because that configuration provided a very lightweight structure. It could be built like a Pratt bridge truss with the wings separated by vertical struts, which were braced by diagonal wires.

Percy Pilcher was the first would-be aviator to develop a gasoline engine to power an airplane. Pilcher, a follower of Lilienthal, made gliding flights from 1896 to 1899. He had planned to use a propeller and his four horsepower engine to power one of his gliders. Unfortunately, he died in a crash in 1899 before he had the opportunity to mount his engine on his glider.

## KEY NUMBER SEVEN • WILBUR'S DEPTH OF UNDERSTANDING

Wilbur Wright was awed by the enormity of the reading and study tasks he had undertaken. He was shocked to learn that there was no such thing as a science of flight. He stated:

*"Contrary to popular opinion we found that men of the very highest standing in the profession of science and invention had attempted to solve the problem…But one by one, they had been compelled to confess themselves beaten, and had discontinued their efforts. In studying their failures we found many points of interest to us. At that time, there was no flying art in the proper sense of the word but only a flying problem. Thousands of men had thought about flying machines and a few had even built machines, which they called flying machines but these were guilty of almost everything except flying. Thousands of pages had been written on the so-called science of flying, but for the most part the ideas set forth, like the designs for machines were mere speculations and probably ninety per cent was false".*

After three months of reading, from June to August 1899, Wilbur Wright had achieved a much better and a more accurate understanding of the field of aeronautics than any one in the world, including

Lilienthal, Langley, and Chanute. These men had spent decades in research and had written books on the subject. How did Wilbur do this? He was a very disciplined person. He established a goal of learning and of achieving deep understanding, and never lost sight of it. He sorted his material as he read, discarding that which was not valid and studying that which was relevant. He also drew Orville into discussions to analyze some of the information to obtain the value of both brothers' thoughts before forming a conclusion. He was a precise engineer analyzing a problem in precise terms, identifying the missing bits of information and then looking for answers to his questions. Other students would have lost themselves in a maze of confusing detail, but not Wilbur.

Some experimenters, including Langley, had come to the field with their own pet ideas and had followed blindly without regard for the experience of their predecessors. Instead, Wilbur carefully studied the work of others, looking for the most fundamental answer to several questions. What does one need to know to learn how to fly? What is known? What needs to be developed? What material is false? Remarkably, Wilbur emerged from his study with answers to these questions. With this uncommon insight, the Wrights raised the field of aeronautics to a new level and brought the invention of a practical airplane to fruition. No other experimenter had taken such an organized approach to the development of a flying machine. Consequently, the deep level of understanding achieved by Wilbur's organized approach to literature search and review was the seventh Key to the Wright Brothers' success.

## Contemporary 1899 Research

Wilbur concluded that the work of Lilienthal, Pilcher, Chanute and Langley was the most useful to him, even though none of their work had produced a full-sized, powered airplane as of 1899. Chanute had not progressed beyond straight-ahead gliding flight. Langley had finally made a successful flight with his steam-powered, unmanned airplane that was 1/3 of full size.

Pilcher and Lilienthal had made significant achievements in gliding flight, and both were planning to mount engines on their gliders. However, both had been killed in glider crashes before their plans could be implemented.

All four of these experimentalists had one major flaw in their work. None had provided for an adequate control system. Lilienthal, Pilcher, and Chanute controlled their craft by swinging their bodies left, right, fore or aft as required, to return their glider to level flight. Langley's unmanned airplane had a very minimal control system; it maintained equilibrium by wing and tail design that insured automatic return to level flight after being disturbed by a gust.

## Lilienthal's Contributions

Wilbur concluded that Lilienthal's work was clearly the best of all. Lilienthal had provided mathematical formulas for computing the lift and drag on airplanes and had determined the lift and drag data for several curved wing shapes. With this information one could design a wing that would lift a given weight at a given air speed, the first step toward designing a successful airplane. Lilienthal also had determined that a curved wing shape produced more lift than one that is flat.

Wilbur thought that Lilienthal was brilliant, innovative, shrewd, and courageous. Still, Wilbur was convinced that Lilienthal was wrong! But Wilbur was concerned about this conclusion. Was he, Wilbur, so far ahead of anyone in the world that he could question such an aeronautical great as Lilienthal? Was everyone in the "aeronautical parade" out of step except Wilbur? He had formed his conclusion because he saw that Lilienthal's work contained two fatal flaws: his methods of control and propulsion.

Consider, first, Lilienthal's body-swinging method of controlling his glider. When a gust of wind would force the right wing tip of Lilienthal's glider upward, he would control the glider by swinging his body to the right. This movement would add weight to the right side and would return the right wing to its the level position, just like one balances a teeter-totter by moving toward the high side to bring it down. Lilienthal is shown making such a correction in Figure 14. Similarly, an up-raised left wing tip would be returned to level by swinging the body to the left. In like manner, when a gust would force the nose of the glider upward, the pilot would swing his body forward. When the tail was forced upward the pilot would swing his body aft and the tail would return to the level flight position.

So what was wrong with Lilienthal's approach? First, it would require a whole series of agile and strenuous acrobatic maneuvers to accomplish this feat. In gusty air this method of control would become an exhausting, if not impossible, physical effort. Further, in both Lilienthal's and Pilcher's cases, this method was not sufficient to prevent their fatal crashes. Finally, this method of control was doomed to become totally inadequate because wings of practical, powered aircraft would have to be much longer than those of Lilienthal's gliders in order to provide sufficient lift to carry the pilot, engine, cargo and passengers. For these larger wings the pilot's weight swinging from the center would not have enough leverage to restore a larger wing to a level flight position.

The second of Lilienthal's fatal flaws was that he planned to propel his plane with bird-like, flapping wing tips powered by an engine. This approach would have been completely impractical, particularly for larger aircraft.

## Wilbur's Summary: Lift, Power and Control

At the end of his literature study, Wilbur summed his conclusions in this way.

*"The difficulties which obstruct the pathways to success in flying machine construction are of three general classes. Such a machine would require wings that would lift it into the air; a power plant to move it forward with sufficient speed so that the air flowing over the wings would generate that lift; and a means of controlling the machine in the air."*

No one before the Wrights had fully recognized that all three of these elements must be addressed when building a successful powered flying machine. Wilbur

had defined the problem in a most succinct manner. The requirements for flight were lift, power and control. Since Lilienthal had done so much to advance the science of lift, and since numerous types of gasoline-fueled engines were being developed, Wilbur determined that only the problem of control remained. No one had solved the control problem, including Lilienthal, who had been studying the problem of flight for 17 years and had been piloting gliders for six years. Thus, Wilbur concluded . . . "the problem of equilibrium (achieved by pilot control) constituted the problem of flight itself." The problem of control would be the driving force of his venture.

Figure 14

Figure 14
Otto Lilienthal attempting to control his glider by swinging his body

# Chapter 4 – The Problem Of Powered Flight

During his study, Wilbur had engaged Orville in many discussions of the flying problem. Thus, by the fall of 1899, Orville was eager to become a partner in the project with his brother.

The Wrights classical manner approach that would make any graduate school professor or research engineer proud. They wrote a statement of the problem that they were trying to solve. This problem statement involved the four forces acting on an airplane in flight - thrust, drag, gravity and lift.

FIGURE 15
The four forces on an airplane in level flight

FIGURE 16
A typical curved cross-sectional shape of a wing showing streamlines flowing around it

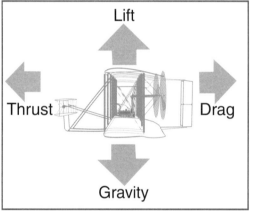

Figure 17

## The Four Forces of Flight

FIGURE 15 depicts the forces that act on an airplane in level, constant-speed flight: thrust, drag, gravity, and lift.

Thrust is the propulsive force provided by the engine and propeller that moves an airplane through the air. By moving the airplane through the air, thrust provides the power required to produce airflow around the wing. The faster the plane flies, the greater the airflow. As airflow increases, more lift is generated. Without airflow around the wing, no lift would be produced. Thus, there is a direct relationship between the available propulsive power, the speed of the airplane and the amount of lift that is generated.

Drag is the force exerted backward on the plane as the air resists the passage of the plane through it.

Drag is the wind resistance one feels when extending an arm out the window of a moving car. Thrust must overcome drag in order for the plane to move forward.

Gravity is the downward-acting force caused by the earth's gravitational attraction that causes all bodies on or near it to fall toward the center of the earth.

Lift is the upward force that raises an airplane off the ground. Lift, is generated, primarily, by the wing. The amount of lift generated is dependent on the shape of the wing's airfoil (the curved shape of a wing, shown in, FIGURE 16. The amount of lift also depends on the wing area, the angle at which the wing is inclined toward the oncoming flow of air and the speed of air flowing around the wing. An airplane's lift must be greater than its weight in order for the airplane to lift off the ground.

In summary, drag and gravity are natural forces inherent in anything moved through the air. Thrust and lift are artificially created forces used to overcome the forces of nature so that an airplane can fly. The engine and propeller combination is designed to produce thrust to overcome drag, as well as to produce enough airflow around the wing to lift the airplane off the ground. The wing is designed to produce enough lift to overcome the gravity

## KEY NUMBER EIGHT •
## THE PROBLEM OF POWERED FLIGHT

Four forces on an airplane in flight were recognized by the Wright brothers, as can be seen in the different problem areas they included their statement of the problem of powered flight:

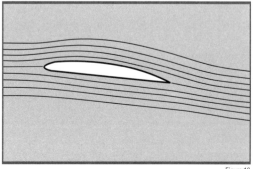

Figure 16

## Problems of Powered Flight

1. Emulate the soaring of birds.
2. Develop a three-axis control system.
3. Solve the secret of lift (and drag).
4. Develop a lightweight structure.
6. Develop an aeronautical propeller.
7. Learn how to fly.
8. Live long enough to learn how to fly.

In writing this statement the brothers had broken down their end objective into a series of sub-problems, each of which addressed a major development effort required for achievement of their end goal. Instead of flailing away at the total problem, it was the Wrights' strategy to solve one problem at a time and proceed systematically from one problem to another until all was achieved. This approach definitely elevated the Wright Brothers above their predecessors and contemporaries. Formulating this statement of the overall problem they set out to solve, breaking the problem down in several smaller steps and then planning to attack these sub-problems one by one was the eighth Key to the Wright Brothers' success. The various elements of this problem statement will be described in the following paragraphs.

## Soaring of Birds

The Wrights soon realized that wing flapping was a dead issue. However, Wilbur's observations of the graceful soaring of large birds and the success of Otto Lilienthal's and Percy Pilcher's gliding experiments convinced him that soaring flight was possible for man. Thus from the beginning the Wrights adopted the concept of a fixed wing design.

## The Secret of Lift and Drag

Wilbur's study demonstrated that a wing having a curved airfoil, shown previously in Figure 16, was more efficient than one having a flat airfoil. Lilienthal and other experimenters also realized that a wing's lift and drag capability varied from one airfoil shape to another. Lilienthal had developed lift and drag data for a number of different airfoil shapes, and he had used existing formulas for computing the lift and drag on a wing. Although the Wrights thought that the secret of lift was in hand, they did not know how to predict the performance of any airfoil shape different from those used by Lilienthal. All they knew was that if they used one of Lilienthal's airfoils they should be able to compute the lift and drag of a wing that incorporated that airfoil.

## Develop a Lightweight Structure and Engine

Since lift opposes weight one can understand why the Wrights and all subsequent aeronautical engineers have worked so diligently to reduce weight; i.e., a reduction in weight reduces the amount of lift required for flight. Since the amount of lift developed by a wing is dependent upon the forward air speed caused by engine power, when weight is reduced, less power is needed for flight. Or looking at the design problem in another way, for a given size of engine, a reduction in weight will permit the plane to fly either longer on one tank of fuel or faster for a shorter period. So, Wilbur exercised great care in designing structures that were both adequately strong for safety and sufficiently light in weight for good flight performance capability. As for the engine, the Wrights believed that they could purchase an engine that would meet their specifications, so they did not make any engine development plans.

## Develop an Aeronautical Propeller

Originally thought to be a minor effort, it was the Wright's plan to use existing technology in engineering handbooks based on marine propellers, and then adapt their findings to meet their airplane needs.

## Learn How to Fly

The Wrights were among the first to realize just how difficult it would be to fly an airplane. Many of the past and contemporary aeronautical experimentalists had required their carriage drivers and laboratory assistants to fly their planes. In effect they said: "…get in and drive it." This way of thinking has since been alluded to as "the chauffeur principle." The Wrights' experience in riding bicycles (which are

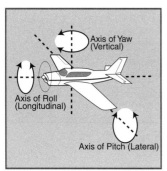

Figure 17

FIGURE 17
The three axes in space about which an airplane must be controlled.

unstable machines) and their recognition that flying would be even more difficult, convinced them there was a need for flight training.

## Live Long Enough to Learn to Fly

Although this was not one of the Wright's listed problem areas; it was a great concern, which they voiced several times. During the previous three years, they had witnessed the deaths of two of the leading pioneers of manned, gliding flight: Lilienthal and Pilcher. They realized that Lilienthal had accumulated only five hours of flight time over a six-year period before he crashed to his death. This certainly was not enough flight experience. Therefore, the Wrights vowed to design and build their aircraft very conservatively and to avoid taking unnecessary risks that would result in serious injury or death.

### Three-Axis Control System

Wilbur Wright realized that an adequate control system was essential for successful flight.

After three weeks of in-depth study, Wilbur had concluded that any object in the air would freely rotate about three axes, each of which was perpendicular to the other two. Thus, there would be a need to control the flying machine about each of these axes.

FIGURE 17 shows a modern airplane with its three axes intersecting at the airplane's center of gravity. Each axis is perpendicular to the other two. One axis passes through the nose and tail of the plane, another extends from one wing tip to the other wing tip, and the third axis passes vertically through the center of gravity of the plane.

Referring to FIGURE 18, rotation about the axis passing through the nose and tail is called roll, just like a ship rolls from side to side in the ocean. Rotation about the axis passing from wing tip to wing tip is called pitch, again just like a ship pitches, nose up and nose down in rough water. Finally, when an airplane rotates from right to left about the vertical axis, its motion is called yaw, like a weather vane turning on its axis.

For full control a pilot causes his plane to pitch nose up or nose down when he wants to climb or dive. When he wants to raise or lower a wing tip, he must be able to cause the plane to roll. And, if he wants to turn the nose of the plane from left to right, he uses his controls to yaw the plane to the left or right. In a similar manner, when a gust of wind hits the plane and disturbs its straight and level flight, the pilot must be ready and able to control the plane's rotation about all three axes simultaneously. Obviously, an airplane requires a much more complex control system than an automobile.

## Two Control Philosophies

Early experimenters thought that man was not quick enough to correct for gust-induced buffeting. The prevailing practice prior to Wilbur's study (and for about ten years later) was to build a self-balancing airplane that would automatically return to its level flight position after being struck by a gust. In addition, two simple controls were used: a vertical, movable rudder in the rear like that on a boat or ship, and a movable horizontal "rudder" in the rear (now called an elevator) to change altitude, just like those used by submarines of that day to rise or descend in the water.

Wilbur Wright had an entirely different concept. His bicycle experience had convinced him that if a man could control an unstable bicycle, he also could control an unstable airplane. Wilbur's concept was to develop a machine that was slightly unstable and which would provide the pilot rapid, complete control about all three-axes.

He was convinced that the prevalent automatic balance concept was dangerous. Heavy turbulence in air could buffet a plane into conditions from which automatic balance systems could not recover. He also believed that existing two axis control systems would not have enough control power to permit the pilot to change course or altitude quickly enough to avoid danger. Wilbur concluded that his control philosophy was superior to all others and that a means for roll control would be the main challenge.

# Roll Control

One day, while Wilbur was closely watching the flight of pigeons and buzzards, he caught the essence of their marvelous control. These are his words.

*"My observation of buzzards leads me to believe that they regain their lateral balance, when partly overturned by a gust of wind, by a torsion of the tips of the wings. If the rear edge of the right wing tip is twisted upward and the left downward the bird becomes an animated windmill and instantly begins a turn (roll to the right), a line from its head to its tail being the axis (of the roll)."*

And that was it! Wilbur had discovered one of nature's secrets of flight. Now, how would he achieve this twisting motion mechanically in an airplane?

He and Orville discussed a number of mechanical roll control methods such as those incorporating wing-length shafts driven by gears that would twist one wing tip one way and the other tip in the opposite direction. However, none of these seemed practical, considering the weight and complexity that was inherent in each system.

## Key Number Nine • Roll Control Discovery

Note that in the following description and in the future, we will be referring to the left and the right wing. By standard definition, the left side (or wing) of an airplane is on the pilot's left when he is facing forward and the right side (or wing) is that side on the pilot's right when facing forward. Thus, when viewing Figures 19 and 20 which are front views of a simulated biplane, the sides are reversed: i.e. left wing is on the reader's right, and the right wing is on the reader's left.

The problem of devising a practical device for controlling roll was continually on Wilbur's mind. Then suddenly one day the light of invention struck. He had just sold a new inner tube for a bicycle tire to a customer and was about to discard the inner tube box when he unconsciously squeezed the box and it twisted in his hand. He immediately tore out the ends of the box,

and saw that he had a model of a biplane wing, as shown in Figure 18.

He used the thumb and forefinger of his left hand to grasp the diagonally opposite top-front and bottom-rear edges of the box while using the thumb and forefinger of his right hand to grasp the top-rear and bottom-front edges. Then, by squeezing with both hands, he deformed the box in the manner shown in Figure 19. In this case, the rear edge of the left wing tip (right tip as you are viewing Figure 19) is raised above the front edge of the left wing tip.

On the other hand, the rear edge of the right wing tip (left tip as you look at Figure 19) is twisted lower than the front edge of the right wing tip. Note that the arrows on the drawing show the direction of the forces caused by the wind striking each twisted wing tip. In view of the directions of the forces in Figure 19, this configuration would cause the left wing tip to drop, (reader's right) the right wing to rise and the airplane to roll to the left. This was the same way Wilbur had noticed the bird's wings had twisted when they wanted to roll to the left.

Conversely, the plane will roll to the right with the right wing tip down (reader's left) and the left wing tip up if you reverse the box's tip twisting, as shown in Figure 20, and the plane would roll to the pilot's right.

The discovery of this method of roll control was the ninth Key to the Wright Brothers' success.

Wilbur was excited but he began to wonder. Was there anything wrong? No! And better yet, Wilbur realized his method would be effective for any aircraft of any size, whereas Lilienthal's body-swinging method extensively limited the size of a plane that could be controlled. Why? Because in Wilbur's method, the forces are applied at the wing tips where one obtains the most rolling leverage. Thus, as airplanes would become heavier and wings would become longer, the forces at the tips would generate even more control torsion than on a shorter-winged vehicle.

No one else had developed a satisfactory method of lateral control. No one had seriously considered wing tip twisting, or "wing warping," as Octave Chanute later

Figure 18

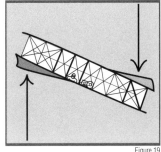
Figure 19

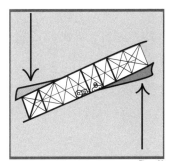
Figure 20

Figure 18
 Box twisted, Wilbur Wright's inspiration for wing warping

Figure 19
 Box twisted to provide roll to the pilot's left

Figure 20
 Box twisted to provide roll to the pilot's right

Figure 21

FIGURE 21
Dr. Paul Garber holding a
model of the Wright 1899 Kite
he had constructed and flown

called it. Wilbur alone had the secret. His was a unique method for roll control. The way was clear for him to advance. One has to marvel that Wilbur invented this system only one month after he had begun his study of the reports he had received from the Smithsonian Institution.

## KEY NUMBER TEN •
## EMPHASIS ON SAFETY

What should be the next step? Build a man-carrying airplane? No! Wilbur was very safety conscious. His development approach based on safety and careful use of resources was to crawl before he stood, stand before he walked, walk before he ran, and run before he flew. The Wright's high degree of safety consciousness in this and other areas was the tenth Key to the Wright Brothers' success. This emphasis on safety assured that the brothers would live long enough to learn how to fly!

Instead of risking life or limb in an untried flying machine to test his wing warping technique, Wilbur decided to proceed more deliberately by first building and flying a kite. His kite project began the first week of July 1899 and was completed by August 6.

### The 1899 Kite

First, Wilbur made a model of the kite he intended to build. It was small, comprised of a paper-covered bamboo framework with cross bracing of thread. When the model was completed, he bent it, twisted it and looked for weaknesses. Satisfied with the

operation and structural integrity of the kite, he designed and built a larger kite with which to test his theory.

FIGURE 21 is a reproduction of the Wright's kite held by its maker, Dr. Paul Edward Garber, the first curator of the Smithsonian Air and Space Museum. He had flown it, and its wing tip is a "little bent"! The kite was a "double decker," as the Wrights used to call it, having a span of five feet and a wing chord (width of the wing) of thirteen inches. It had a pine framework covered with fabric that had been painted with shellac to reduce the fabric's porosity. FIGURE 22 contains five sketches of the kite. FIGURE 22b shows that the front and rear rows of vertical struts were braced diagonally. To facilitate wing warping, no bracing was provided between the front and rear struts. The joints at the intersection of the vertical struts and the horizontal wing spars were hinged to permit the wings to pivot forward and rearward relative to each other, as seen in FIGURE 22d. The rear-mounted horizontal stabilizer (the Wrights called it a "rear rudder") was attached rigidly to the middle of the rear, center vertical strut, as in FIGURE 22a. A small weight was placed on the bottom wing at center span, about one-quarter of the way back from the leading edge of the wing, FIGURES 22a and 22b.

The kite's control system included four lines fastened to the upper and lower tips of the wing and led to two vertical control sticks which were to be held in both hands by the operator. With this system the kite could be controlled in both pitch (nose up or down) and roll (left wing up and right wing down, and conversely). Notice that the lines were crossed.

Pitch control was achieved by pointing the tips of both control sticks either toward or away from the kite. By pointing the tops of both sticks toward the kite, the upper wing would move ahead of the lower wing as in FIGURE 22d. Then, the kite would pitch nose up as in FIGURE 23. By moving the tops of the two control sticks toward the operator, the upper and lower wings would assume the positions in FIGURE 22e, and the kite would pitch nose down and dive, as in FIGURE 24.

Roll control was achieved by pointing the top of the stick in one hand toward the kite and the top of the stick in the other hand away from the kite, in an X-configuration. In Figure 22c the operator had pointed the top of the left hand stick toward the kite and the top of the control stick in his right hand away from the kite. In this case the wing tip on the operator's right would be twisted with the leading edge down relative to the trailing edge, and the wing tip on the operator's left would be twisted with the leading edge up relative to the trailing edge. With a twist in this manner, the kite would roll to the reader's right, as shown in Figure 25. By reversing the movement of the control sticks, the roll would be made to the reader's left, as in Figure 26.

## Kite Test Flight

Walter and John Reinieger and a few other neighborhood boys accompanied Wilbur as he went to test his kite. The kite flew spectacularly. Best of all, the wing warping controls worked perfectly as Wilbur directed the kite in smooth turns from left to right and back again. When he operated the pitch control for a climb the kite responded quickly to his command. However, when he directed the kite to head nose-downward that was another story. As the kite pitched downward it lost altitude so rapidly that the control lines became slack, causing Wilbur to lose control. As the kite swooped down, the boys had to dive to the ground to avoid being hit. Fortunately,

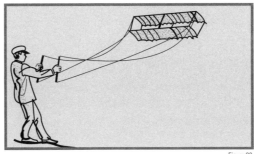
Figure 23

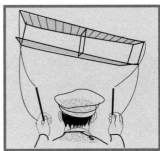
Figure 25

the kite recovered low to the ground and climbed to a safer altitude

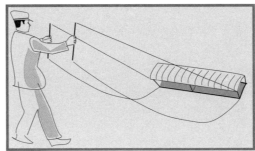
Figure 24

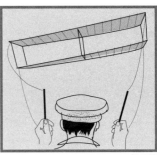
Figure 26

Wilbur was so elated with the results that he went to the campsite where Orville, Katharine and friends were on a chaperoned camp out. He took Orville aside, and with wild gesticulations and an excited voice told Orville that the wing warping control worked to perfection. It seemed that their problem had been solved.

But one problem remained. Why did the kite dive so out of control and what could be done to correct for this unacceptable flying behavior?

Figure 22
   Sketchs of Wright 1899 Kite

Figure 23
   Wright kite being controlled in pitch to make it climb.

Figure 24
   Wright kite being controlled in pitch to make it dive.

Figure 25
   Wright kite being controlled in roll to the reader's right.

Figure 26
   Wright kite being controlled in roll to the reader's left.

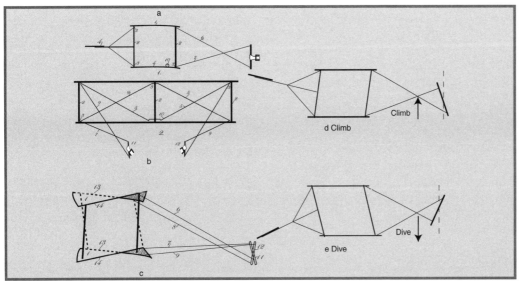
Figure 22

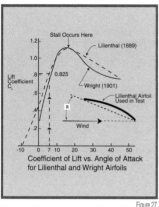

Figure 27

FIGURE 27
  Lilienthal's and Wrights'
coefficient of lift vs.
Angle of Attack graph for
airfoil shaped like the
arc of a circle.

# Chapter 5 – The 1900 Glider

Wilbur Wright was excited and ready for a new challenge now that he had demonstrated that his control concepts operated satisfactorily. He was now ready to take the next step: build a man-carrying machine and learn to fly it. Obviously his family was very concerned about Wilbur's safety and would have preferred that he confine his aeronautical activities to study and kite flying.

In response to these concerns, Wilbur wrote:

*"Now there are two ways of learning how to ride a fractious horse. One is to get on him and learn by actual practice how each motion and trick may be best met; the other is to sit on a fence and watch the beast a while, and then retire to the house and at leisure figure out the best way of overcoming his jumps and kicks. The latter system is the safest; but the former, on the whole, turns out the larger proportion of good riders. It is very much the same in learning to ride a flying machine; if you are looking for perfect safety, you will do well to sit on the fence and watch the birds; but if you really wish to learn, you must mount a machine and become acquainted with its tricks by actual trial."*

## 1900 Glider Design

Wilbur selected a glider for his first man-carrying machine because it would be simpler, less expensive and safer than a powered machine. It also would be quite satisfactory for testing his control concepts.

The Wrights' first glider was not the product of guesswork. Instead, Wilbur selected the best information he could obtain and then added his own judgment to literally design his glider by use of mathematical and engineering principles.

Wilbur's first question was: What is the required wing size? The first design task was to determine the appropriate wing area that would provide sufficient lift to raise the machine and a pilot off the ground. Wilbur had obtained the following formula for wing lift from his reading:

$$L = K \times C_L \times S \times V^2$$

This formula simply states that Lift increases or decreases directly as the coefficient of lift, $C_L$; the Smeaton coefficient of pressure, K; the wing area, S; and the airspeed squared, $V^2$. The terms K and $C_L$ will be defined in the following paragraphs.

The value K in the above formula is the Smeaton coefficient of air pressure, which was described in Chapter 3. K is essentially a measurement of the density of air (i.e. the total weight of air molecules that occupy one cubic foot of space). Its constant value of 0.005 was attributed to the work of John Smeaton, a British engineer. This number had been used, with few exceptions, since its introduction in 1759 until the early 1900's.

The coefficient of lift, $C_L$, is not just one number. Instead, it is a table of numbers determined experimentally. The table consists of values of $C_L$ vs. angle of attack; i.e. the angle at which the wing is inclined with respect to the oncoming air during wing airfoil tests. Each wing airfoil shape has its own unique set of $C_L$ vs. angle of attack values. During an airfoil test, the wing is inclined at a one angle of attack and the lift is measured. Using the preceding lift equation, the corresponding value of $C_L$ for that angle of attack is computed. Then the wing is inclined at a different angle of attack and the new value of $C_L$ is computed. The test continues until all values of $C_L$ over the desired range of angles of attack have been obtained. Finally, the computed values of $C_L$ vs. angle of attack are recorded in tabular form and then are plotted in a graphical form.

FIGURE 27 is one of these graphs of tabular airfoil lift coefficient data obtained from experiments conducted by Otto Lilienthal in 1889 and by the Wrights in 1901. This graph shows how coefficient of lift, $C_L$, increases or decreases as the angle of attack increases or decreases. FIGURE 27 presents the graphs for two airfoils: the Lilienthal 1889 airfoil and the Wright 1901 airfoil, which were slightly different.

Lilienthal had experimentally determined this set of data by measuring the lift and drag on a wing of a known area, which was inclined at various angles of attack, as described previously. He obtained the moving air, not in a wind tunnel, but by mounting the wing at a given angle of attack on the end of a rapidly rotating horizontal pole in order to cause moving air to blow on the wing model. After measuring the lift on the wing, he computed $C_L$ using the preceding formula. For example, from one of these computations he determined that $C_L$ was equal to 0.825 when the wing's angle of attack was seven degrees. He then repeated his test for various angles of attack.

Wilbur Wright had calculated that the his glider wing had to carry 190 lb – his weight of 140 pounds plus the glider's estimated weight of 50 pounds. Using the above formula and the $C_L$ data in FIGURE 27, Wilbur determined that a wing having an area of 200 square feet would support his piloted glider when flying at a speed of 15 miles per hour at a wing angle of attack of about seven degrees.

Next, Wilbur selected Octave Chanute's wing design concept, shown in FIGURE 28, i.e. two parallel wings separated by vertical struts and braced by diagonal wires. He realized this type of wing, which looked like a Pratt bridge truss, would provide a marked improvement over Lilienthal's monoplane design because it was stronger and would be lighter in weight. Since two wings would be used, each wing would have an area of 100 square feet (1/2 of the total computed 200 square foot area).

By first selecting a chord, i.e. width, dimension of 5 feet for each wing, and using the 100 square foot area of each wing, he calculated the wing span:

**Area = Span x Chord**

or,

$$\text{Span} = \frac{\text{Area}}{\text{Chord}}$$

or,

$$\text{span} = \frac{100}{5} = 20 \text{ ft.}$$

Now, what airfoil shape should Wilbur use? He had been examining a Lilienthal airfoil that was shaped like the arc of a circle, with its high point at the center of the chord. See FIGURE 29. Lilienthal's

maximum airfoil camber (maximum height) was 1/12 of the chord, i.e. 5 inches high. Wilbur thought this shape would create too much drag and would cause other aerodynamic control problems. So, he decided to make two changes; he selected a maximum camber of 1/22 of the chord, i.e. 2.7 inches high and chose to place the point of maximum camber only about four inches behind the leading edge, as shown in FIGURE 29.

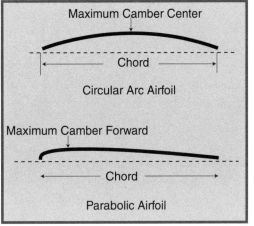

Figure 29

The Wrights' experience in bicycle racing had taught them that a rider should lean low over the handlebars in order to reduce his frontal area and hence reduce the drag that would have to be overcome by pedaling. For that reason Wilbur planned for the pilot to assume a prone position rather than the vertical position that Otto Lilienthal, Percy Pilcher, and Octave Chanute's assistants had assumed when they hung from their gliders. Wilbur calculated that a prone pilot's frontal area would be only 1/10 that experienced by other glider experimentalists, and that total drag could be reduced by 50 percent.

One last problem remained. What could be done to correct the uncontrollable diving tendency of Wilbur's kite?

The brothers decided to counteract this problem by placing a controllable, horizontal stabilizer in front of the wing. They called this a front rudder; however today we call this front-mounted horizontal stabilizer a canard, which comes from the French word for duck. The word canard is used, possibly, because the small control surface in front followed by an airplane's

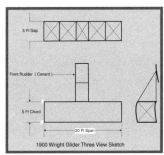

Figure 28

FIGURE 28
Canard Biplane configuration selected by Wilbur Wright for 1900 Glider.

FIGURE 29
Comparison of Lilienthal's Circular Arc Airfoil with the Parabolic Airfoil selected for the Wright 1900 Glider.

narrow, front structural framework and then by the large central area, looks somewhat like a duck in flight with its neck outstretched. The Wrights used a canard on all of their subsequent gliders and flying machines until about 1910. By turning the front rudder up or down, the pilot was able to control a machine's pitch quickly and accurately. See the front rudder (canard) in Figure 28.

## Key Number Eleven • A Canard For Pitch Control

The adoption of a canard became a critically important decision. By a happy accident of design, the canard dramatically changed for the better the stall characteristics of both Wright gliders and their powered airplanes. A comparison of the stalling characteristics of a Wright canard aircraft and that of a conventional airplane having the tail in the rear is shown in Figure 30.

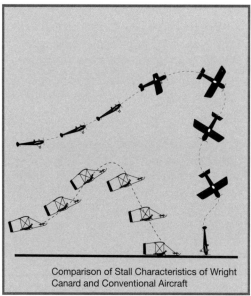

Comparison of Stall Characteristics of Wright Canard and Conventional Aircraft

Figure 30

**Figure 30**
Comparison of stall characteristics of airplanes having a canard horizontal tail and a conventional horizontal tail.

Figure 27 shows the stall point, where the value of $C_L$ drops rapidly beyond its peak value. The stall occurs when the angle of attack becomes so large that the airflow over the top of the wing becomes exceptionally turbulent and separates from the wing surface. This type of flow causes a dramatic loss of lift. When a conventional airplane stalls, it can roll to the left or right

and then fall into a dreaded tailspin. A stall near the ground with a conventional airplane design almost always would cause a fatal accident. Lilienthal and Pilcher both were killed when their gliders stalled about 50 feet in the air.

On the contrary, Figure 30 indicates that the Wright canard was a much safer design. As a properly designed canard stalls, it will automatically recover and remain in a nearly level position descending like a flat, falling leaf. I can vouch for this performance trait because the radio controlled Wright Flyer, which I built, recovered from stalls in exactly this same manner. [Author]

Why was this canard configuration so important? On several occasions while flying, both Wilbur and Orville were struck by gusts of wind that carried their flying machines to heights well beyond their planned safety limits, some as high as 50 feet. In each of these circumstances, they experienced only moderately hard, flat landings from which they escaped with only minor bruises. The adoption of the canard configuration is the eleventh Key to the Wright Brothers' success, because both of them, had they used a rear-mounted horizontal stabilizer, surely would have died in crashes that actually occurred early in their flying careers.

The Wrights did not use a vertical tail on the 1900 glider.

### Basic Design Completed

The foregoing design accomplishments were even more remarkable when one realizes that Wilbur had essentially completed all of the preceding design work only three months after he had written his letter to the Smithsonian asking for literature on aeronautics. Very little information is recorded concerning Wilbur's glider developments until late summer of 1900. Why? Because Orville and he needed to spend all of their time during the ensuing fall, winter and spring manufacturing new bicycles for the next year's trade. Can't you feel Wilbur's frustration? With his dream of flying now nearing reality, he had to assign second priority to the flying machine and concentrate on making a living and earning the money required for their aeronautical experiments. We can only surmise that there were many late nights spent making computations, designing mechanisms, drawing plans, selecting materials and looking for suitable places to conduct their flight tests.

No one knows exactly when construction actually began on the glider. In a letter to Chanute dated August 10, 1900, Wilbur said he was about ready to begin construction, and asked where in Chicago he could obtain suitable spruce lumber for its framework. And then, surprisingly, on September 5, Katharine wrote to the Bishop saying:

*"We are in an uproar seeing Will off (to Kitty Hawk, North Carolina). I don't think he will be reckless. If they can arrange it, Orv will go down as soon as Will has the machine ready."*

So in only about 25 days, Wilbur and Orville had found sources of materials, purchased them, and had built their glider (including sewing the white sateen fabric used to cover the wings and canard). This short construction time was an extraordinary accomplishment, especially for two men making their first man-carrying glider.

Now, the Wright Brothers were ready to "mount the machine and become acquainted with its tricks."

## Test Site Selection

Where shall we test it? Wilbur had designed his glider and this was the next logical question. The answer was presented in responses to letters Wilbur had written to the Weather Bureau and Octave Chanute, the famous aeronautical authority: test at Kitty Hawk, North Carolina! It had wide-open plains, high hills, soft sand and strong, steady winds of the magnitude that Wilbur required. It also was free of trees and was so isolated that there would be no interference by curious onlookers.

However, it was the warm, unsolicited letter from Postmaster William J. Tate, a prominent citizen of Kitty Hawk that clinched the decision. Tate strongly corroborated the Weather Bureau's recommendation and welcomed the Wrights saying:

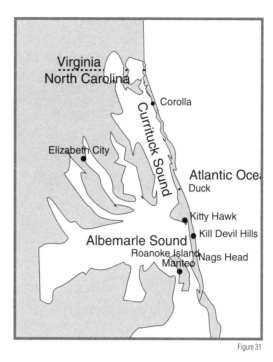

*"You will find us a hospitable people when you come among us."*

Tate also invited the brothers to room and board at his home until they were settled in their own quarters. He suggested that they bring a tent.

## Journey to Kitty Hawk

Wilbur left for the Outer Banks with the crated glider parts on September 6, 1900, at 6:30 p.m. Orville would join him when the glider was nearly ready to be flown.

The trip to Kitty Hawk was an arduous undertaking. His trip included five segments: the Big Four train from Dayton to Cincinnati; the Chesapeake and Ohio to Old Point Comfort, Virginia; the steamship "Pennsylvania" across Chesapeake Bay to Norfolk, Virginia; another train to Elizabeth City, North Carolina; and a sailboat across Albemarle Sound to Kitty Hawk. FIGURE 31 shows the Outer Banks area, the location of Kitty Hawk and the Wright test site at Kill Devil Hills.

In order to minimize his baggage, Wilbur had decided to purchase the four, 18-foot, spruce wing spars in Norfolk. However, when he arrived he learned that 16-foot, white pine spars were the longest that he could obtain. Disappointed, he bought them anyway, knowing that his wings would

Figure 32 Courtesy of Special Collections and Archives, Wright State University

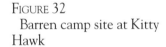

FIGURE 32
  Barren camp site at Kitty
Hawk

FIGURE 33
  Wilbur Wright on KP
duty at Kitty Hawk

be two feet shorter than required and that higher wind speeds would be needed to carry a man aloft.

After four days of searching for a boat at Elizabeth City, Wilbur hired an old fisherman, Israel Perry, to take him to Kitty Hawk. Wilbur was shocked at the condition of the boat. It was rat-infested, filthy, and was equipped with deteriorating ropes, sails and rudder. In view of the difficulty he had in finding any boat, Wilbur went with Perry in spite of the boat's condition.

The trip across Albemarle Sound was a hair-raising event. Wilbur and the other two men on the boat were nearly drowned after the boat became damaged in a vicious storm. Wilbur had embarked the early evening of September 11. Yet, it wasn't until late the night of September 12 that the boat arrived at Kitty Hawk because they had to seek overnight haven from the storm

in a sheltered cove and make repairs to the damaged boat. Arriving too late to proceed to the Tate's home, Wilbur had to stay on board another night. Mindful of his health, he ate nothing offered by Israel Perry, and slept topside to avoid the filth below. The only thing he had eaten for two days was a small jar of jam that Katharine had tucked into his trunk. He was famished.

Finally, on the morning of September 13, seven days after leaving Dayton, a boy escorted Wilbur to William Tate's home where he was given a good breakfast.

Wilbur eagerly began assembly of his glider under an awning in Tate's yard. Assembly was nearly completed when Orville arrived on September 28. The two men stayed with the Tates until October 3 because Wilbur had to re-cut and sew his wing coverings, made of white French sateen, to accommodate the two-foot reduction in wingspan caused by the shorter-than-planned spars. Mrs. Tate kindly offered him the use of her precious sewing machine to make the alterations. The fabric's pores were not sealed (painted or doped) a design practice that the Wright's followed on all of their gliders and powered machines until 1910.

Kitty Hawk was indeed a barren, isolated place. There was practically nothing but sand everywhere one looked. There were no hotels, motels, hardware stores, lumberyards, laundries or restaurants, and very little in the way of groceries. They had to bring with them all they needed or order supplies, which were brought in by boat from Elizabeth City.

Orville had brought a large, 12 ft. x 24 ft. tent, cots, a gas stove and various food supplies and equipment. They set up their tent about a half-mile from the Tate's home. They even dug a well to provide their own supply of water for cooking, drinking and washing. FIGURE 32 shows the desolate Wright campsite at Kitty Hawk and FIGURE 33 shows Wilbur doing KP chores after a meal. Orville did the cooking.

Figure 33 Courtesy of Special Collections and Archives, Wright State University

## KEY NUMBER TWELVE • VALIDATE THEORY WITH TEST DATA

It appears that Wilbur did all of the flying in 1900 and 1901, probably to protect his younger brother from injury. As we will learn later, Orville first flew in 1902 and also became an outstanding pilot.

The 1900 glider was flown in several ways: as an unmanned, tethered kite; a manned tethered glider; a free flying, unmanned glider; and finally as a manned free flying glider.

Most of the time the glider was flown as an unmanned kite, as in FIGURE 34, because the wind speed had to be dangerously high in order for the smaller-than-planned wings to lift the manned glider. Usually, Wilbur and Orville held tethers attached to the lower wing tips and payed out the lines as the wind lifted the glider from the ground. At other times Bill Tate and his half-brother, Dan Tate, helped. Different weights in the form of log chains were attached to test the glider's lifting power.

While flying the glider as a kite, the brothers measured the glider pull on the tethers with a spring scale. They also obtained wind speed measurements from a hand held anemometer lent to them by personnel from the nearby weather station and measured the angle at which the wing was inclined toward the wind. Since they also knew the area of the wings and the glider weight, they were able to calculate for each flight the lift and drag forces.

Finally, they compared their experimental measurements and calculations with the theoretical values they had gleaned from the technical literature. This comparison enabled them to determine whether or not their experimental data agreed with accepted theory and the experimental results of other aeronauts. If the data matched, they knew their results were technically sound. If the data did not match, either something was wrong with the experiment or there was some unknown theoretical factor about which they did not know.

Unfortunately, the latter was the case. The experimental lift and drag data

Figure 34 Courtesy of Special Collections and Archives, Wright State University

FIGURE 34
1900 Glider flown like a kite to measure lift and drag forces

were only about one-half as large as the accepted information in the literature. They corroborated this finding time and time again by adding different amounts of log chain weight. Something was terribly wrong. Was Otto Lilienthal's airfoil data wrong? Were the lift and a drag equation used by everyone else wrong or was there another flaw?

These kite flying sessions distinguished the Wright Brothers from other aeronautical experimentalists. Their approach to measuring flight data and comparing their experimental results with accepted theory was the twelfth Key to the Wright Brothers' success. This experimental approach is the same as used in classical research programs today: validate theory by comparing it with experimental data!

On one occasion Wilbur went aloft on a tethered glider when the wind was 25 miles per hour. By carefully actuating the wing warping and the front rudder pitch controls, he verified that they operated as expected. When the glider began to climb higher than 15 feet, he called to the men to lower him immediately to the ground for safety reasons. On another occasion Dan Tate's son, Tom, was flown aloft on the tethered glider because he weighed much less than Wilbur and hence required less wind speed to lift him.

On the day before they had planned to return to Dayton, they took the glider to the big hill for some manned flight experience. They first flew the glider with no one aboard by running with it to the brow of the hill and releasing it into the wind. It would fly for a short distance, and then stall or dive. Seeing that it was quite docile in flight, they decided it was safe to attempt a manned free flight.

Wilbur lay down on the lower wing and Orville and Dan Tate grasped opposite lower wing struts. When all were ready, the two wingmen, carrying the glider, ran down the hill into the wind and released it after flying speed had been attained.

Wilbur was flying freely!

To his delight, Wilbur learned that the glider's pitch (nose up, nose down) was easy to control, and that he could bring it in for a gentle, smooth landing. His flights were about 300 to 400 feet in length and lasted about 15 seconds per flight. For safety reasons Wilbur kept the glider no more than 5 feet above ground and resisted temptation to fly higher even though the glider seemed to encourage him to do so.

## Home to Dayton

They broke camp to return to Dayton on October 23 and abandoned the glider where it lay after the last flight. They had no further use for it. Shortly after they had left, Mrs. Tate was seen walking purposefully toward the glider carrying a pair of scissors. She didn't view the French sateen as glider covering. She saw it as new dresses for her two youngest daughters!

In evaluating the tests of 1900, Wilbur said:

> "We were very much pleased with the general results of the trip, for setting out as we did, with almost revolutionary theories on many points, and an entirely untried form of machine, we considered it quite a point to be able to return without having our pet theories completely knocked in the head by the hard logic of experience, and our own brains dashed out in the bargain,"

Yet, above all, Wilbur was elated. He had flown! He had experienced the thrill of soaring through the air.

However, there were some very discouraging results for which they had no explanation. The measured lift and drag of their glider was only about one-half of that expected when compared to Lilienthal's tabular data and the standard lift equation. To get the desired lift, they would have to fly at a wing angle of attack of twenty degrees instead of three degrees as Lilienthal's data had indicated. In spite of their careful planning, something was terribly wrong.

Also, they were very disappointed that they had not achieved a major objective of the trip: the accumulation of many hours of flying time and piloting experience. In all, Wilbur had logged only two minutes of piloting during a month's time.

They arrived in Dayton with fervor to push on but were troubled by major questions. They hoped to find answers to these questions with their new glider, which they planned to test at the Outer Banks the following year.

# Chapter 6 –
# The Largest Glider in the World

The Year 1901 dawned with the brothers eagerly looking forward to building a larger glider and testing it at Kitty Hawk.

## The 1901 Glider

Their new glider would have a span of 22 feet, a chord of 7 feet and would be 14 feet long. With a 290 square foot wing area and an 18 square foot front rudder area for control, this would be the largest glider that had ever had been built. They also decided to nearly double the depth of the wing's airfoil at its widest point to 5 inches from the 2.7 inches of the 1900 glider wing. This larger curvature would be more like that used by Otto Lilienthal for his glider. This change was made because the Wrights thought that the poorer-than-expected performance obtained from their 1900 tests might have been a result of their decision to deviate from Lilienthal's airfoil shape. A picture of the 1901 glider with Orville standing beside it is shown in FIGURE 35.

Wilbur had continued his correspondence with Octave Chanute. Chanute was becoming more and more interested in what these "Dayton boys" were doing. More importantly, he asked to visit the brothers in Dayton, and he was invited with enthusiasm. He also offered to loan them his anemometer for measuring wind speed at the test site. The Wrights gratefully accepted.

After this visit, Chanute asked if he and two of his assistants, Edward C. Huffaker and George A. Spratt, could visit the brothers at Kitty Hawk during the 1901 tests. Chanute wanted to test a glider that he had commissioned Huffaker to build for him. In this way, Chanute could get the Wright's help in testing his glider; also he could give a little help in return. The Wrights welcomed their visit.

## Back to Kitty Hawk, 1901

Wilbur and Orville left Dayton at 6:00 p.m. on July 7, 1901 for their second trip

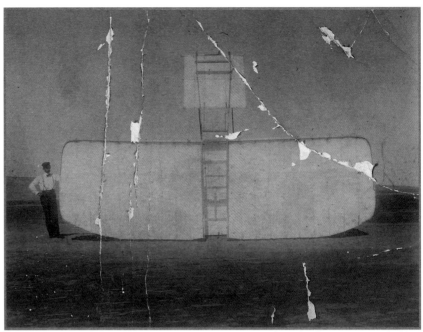

Figure 35 Courtesy of Special Collections and Archives, Wright State University

to Kitty Hawk. This time they set up camp about four miles south of Kitty Hawk at Kill Devil Hills, so that they would not have to lug their glider long distances to the hills. This is Wilbur's description of the site:

" *The practice ground at the Kill Devil Hills consists of a level plain of bare sand, from which rises a group of detached hills or mounds formed of sand heaped up by the winds. These hills are constantly changing in height and slope according to the direction and force of the prevailing winds. The three which we use for gliding experiments are known as the Big Hill, the Little Hill and the West Hill, and have heights of 100 feet, 30 feet and 60 feet, respectively.*"

This year, in addition to their tent, they built a 16-foot by 25-foot wooden shed. This was their workshop and glider storage facility. To make it more useful and comfortable, full-width doors at both ends were hinged at the top so they could be raised like awnings. This feature simplified the stowage and withdrawal of the glider and provided shade and ventilation during hot weather. FIGURE 36 is a picture of the 1901 campsite arrangement showing the visitors, who arrived on about July 25th. Huffaker was immediately impressed with the quality of the Wrights' workmanship.

FIGURE 35

The 1901 Glider with Orville standing alongside. This glider with a wingspan of 22 feet and a chord of 7 feet was the largest glider that had ever been made.

Figure 36 Courtesy of Special Collections and Archives, Wright State University

FIGURE 36
1901 campsite showing new work shed/ storage building and tent living quarters.

FIGURE 37
Launching the 1901 glider.

FIGURE 38
Wilbur flying the 1901 glider.

FIGURE 39
Wilbur landing the 1901 glider. Note Orville's footprints running up to him possibly asking: "are you all right"? Then he runs back to take the picture.

Two days later, Wilbur made 17 glides. He was shocked. The machine would hardly fly, and it was nearly unmanageable. The only good thing he could report was that, on two occasions, the glider did not crash after it had zoomed to a height of about 40 feet and then stalled. Instead, it came in for harmless, flat landings. This performance probably saved Wilbur's life. Now, at least, the Wrights knew the machine was much safer than Lilienthal's.

They flew the unmanned glider as a kite and took measurements to learn why its performance was so poor. The results were very disappointing. As during the 1900 tests, the lift and drag were significantly lower than Lilienthal's data had indicated and the drag was considerably greater than that of the 1900 glider. They also flew the glider tethered with a man aboard and continued to find that the glider was nearly

impossible to control in pitch (nose up, nose down).

To study the problem further, they took the glider apart and tested one wing alone by attaching spring scales to the two wing-tip tethers and measuring the forces at different wind speeds. From these tests, they determined that the additional curvature they had built into the wing had caused the control instability. A thinner wing would have less drag, and it would not perform so erratically in pitch. They also found that the leading edge of the wing was much blunter than that of their 1900 glider.

It was then that the Wrights' engineering genius and ingenuity came to the fore. They added a third wing spar between the front and rear spar that stretched from tip to tip. Then they installed a trussing arrangement that would press down on the third spar at several locations along the wingspan in order to flatten the high point of the ribs. In this way they reduced the curvature to the same thickness they had used on their 1900 glider. They also changed the shape of the leading edge so that it was more streamlined.

The result of these changes was remarkable. With two men launching the glider, as shown in FIGURE 37, the glider flew very well. FIGURE 38 shows Wilbur in free flight and FIGURE 39 shows him after making a smooth landing. Glides of up to 390 feet were made with perfect pitch control. Huffaker and the others were excited by this performance. By this time Huffaker had given up on the glider he had built for Chanute. It didn't fly and it fell apart, so he discarded it and spent his time watching the Wrights' work.

But Wilbur was not content to hold the world's gliding records. He was dismayed and very discouraged by the fact that, even after the modifications, the measured lift and drag forces were markedly different from existing theory.

Then another problem arose while attempting to turn the glider with the wing warping mechanism. For example, when Wilbur tried to turn the glider to the left with his wing warping control, the glider would start to turn to the left as expected. Then, halfway through the turn, it would

Figure 37 Courtesy of Special Collections and Archives, Wright State University

start to turn to the right without any control change by Wilbur. In other words, the right wing, which would be raised higher than the left wing in a left turn, would start to slow down and the lowered left wing would start to speed up causing the machine to turn to the right. This maneuver made it very difficult for Wilbur to hold while the glider spiraled around in the opposite direction from that desired while tilted in the wrong direction, as the artist depicts in FIGURE 44. The Wrights called this phenomenon well-digging. The same phenomenon occurred for a right turn.

The Wrights had no explanation for this unexpected performance.

Then, a storm blew in; it rained for four days, the visitors left and the brothers decided to go home. They were very discouraged because their best ideas and knowledge had been decimated. They were not getting the lift and drag that the literature said they should be getting and now even their ingenious idea for control (wing warping) that had been proven sound by their early work was not working. The only good they could infer from their 1901 experiments was that the structure was light and strong and that when the glider stalled, it landed safely instead of going into a deadly, spinning nosedive.

Realizing that it was a waste of time to continue testing at this time of the year, they broke camp earlier than expected and returned to Dayton.

## Major Decision

The Wrights felt very dejected when they left Kitty Hawk in August of 1901. Neither man said much on the return trip to Dayton. Wilbur was very uneasy. What he had thought to be solid ground had suddenly become quicksand. He didn't know what to believe and what not to believe. The brothers' feelings at that time were best expressed by Wilbur's recollection.

*"We doubted that we would ever resume our experiments. Although we had broken the record for distance in gliding, and although Mr. (Octave) Chanute, who was present at that time, assured us that our results were better than had ever before*

Figure 38 Courtesy of Special Collections and Archives, Wright State University

Figure 39 Courtesy of Special Collections and Archives, Wright State University

Figure 40

FIGURE 40

The unexpected "Well Digging" performance" of the 1901 Glider.

been attained, yet when we looked at the time and money which we had expended, and considered the progress made and the distance yet to go we considered our experiments a failure. At this time I made the prediction that men would sometime fly, but that it would not be within our lifetime."

Orville's version of what Wilbur had said was even more pessimistic:

"…not within a thousand years would man ever fly!"

When they arrived home, Katharine was surprised to see them. In a letter to her father she said:

"…the boys walked in unexpectedly on Thursday morning…. [They] haven't had much to say about flying. …Will is sick with a cold or he would have written to you before this…"

Wilbur and Orville were ready to give up their quest, but Katharine encouraged them, and Chanute arranged for Wilbur to present a paper before the prestigious Western Society of Engineers in Chicago. Wilbur was reluctant to accept, but Katharine "nagged him into going," saying that meeting all these prestigious men would be good for him. Wilbur decided to go. As he was getting ready to leave for Chicago, Katharine wrote to her father:

"A week from today is Wilbur's speech at Chicago. We asked him whether it was to be witty or scientific and he said he thought it would be pathetic before he got through with it!"

Getting Wilbur ready for his trip to Chicago was a family affair. In a letter to her father, Katharine wrote:

"We had a picnic getting Will off to Chicago. Orv offered all his clothes, so off went 'Ullam' (her pet name for Wilbur) arrayed in Orv's shirt, collars, cuffs, cuff links and overcoat. We discovered that to some extent, 'clothes do make the man' for you never saw Will look so swell".

Surprisingly to Wilbur, an audience of about 70 members of the Society and their wives enthusiastically received the paper. In fact, the paper was well written, thorough and demonstrated that the Wrights had learned much more about "flying machines" than their contemporaries

The publication of Wilbur's address that followed was hailed throughout the world as "the Book of Genesis of the 20th century, the Bible of Aeronautics". In his book, Wilbur and Orville, Fred Howard, wrote:

"The ripples created by his (Wilbur's) speech fanned out beyond Chicago as soon as it had been printed in the December issue of the Society's journal under the modest title, 'Some Aeronautical Experiments'. Chanute ordered three hundred reprints struck off, half of which he distributed broadside to his aeronautical, scientific and engineering correspondents throughout the world."

During the next few months the paper was reported, abstracted, or printed in full in Engineering Magazine, Scientific American, Fielden's Magazine, and the British Automotor Journal. The magazine Flying got maximum mileage out of the speech by publishing it as a four-part series over a period of ten months. It was included in its entirety in the 1902 annual report of the Smithsonian Institution, which also published it in the same pamphlet format as the four publications sent to the Wrights in 1899.

Such was the outcome of the "pathetic speech" that Wilbur promised his family before going to Chicago. Just think! In a little more than two years the Wright's had risen from the ranks of people interested in flight to persons who had advanced the state of the art to record levels!

Thanks to Chanute, the Wright's interest in flying machines had been rekindled and they began to re-examine other experimenters' information that they had been taking for granted.

Since the Wrights had been the first people to measure the lift of a full-sized flying machine, they were the first to realize that the current state of technology greatly over-predicted the amount of lift and drag that would be realized from a wing of a given design. Everything they had done to date had been based on faith in Otto Lilienthal's work. If they were to change the wing curvature shape used by Lilienthal, they had no idea how the change would affect the wing lift and drag. And finally, they were dumbfounded that their wing-warping system for controlling the glider in roll was not working, even though their kite experiments had indicated that the concept was sound.

## KEY NUMBER THIRTEEN • DECISION TO DEVELOP THEIR OWN DATA

At this time, the Wrights made a major decision. They realized that whenever they used information and data generated by someone else, their gliders did not fly as predicted. They concluded that:

*"…the calculations upon which all flying machines had been based were unreliable and…every experiment was simply groping in the dark…We cast it all aside and decided to rely entirely upon our own investigations."*

In this way the Wrights would know how their set of data was generated and they could be confident of its accuracy and applicability.

What a monumental decision this was! Here were two men who didn't have formal education beyond high school deciding that all of the famous learned and respected men of the past and present had been wrong. Further it would seem unimaginable that

they, the Wright Brothers, would be able to develop information which would be better than that accepted throughout the world.

Their father's early training enabled this decision. The Bishop had taught them that if they honestly believed that something was right, they should hold fast to their convictions and proceed without letting any one dissuade them. And so they went their own way!

Making and acting on this decision was the thirteenth and possibly the most important Key to the Wright Brothers' success.

Figure 42 Courtesy of Carillon Historical Park Archives

FIGURE 41
Wright glider in flight with lift and drag equations.

FIGURE 42
Bicycle wind tunnel showing horizontal wheel mounted on the handlebars.

FIGURE 43
Top view of the horizontal bicycle wheel showing positions of the curved and flat airfoils on the wheel rim.

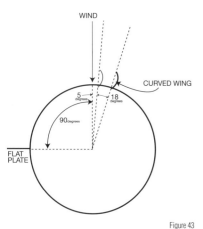

Figure 43

# Chapter 7 – Going Their Own Way

Now that they had decided not to accept time-honored data, the Wrights realized that they had to answer three basic questions. What are the correct values for $C_L$ and $C_D$ (lift and drag coefficients in FIGURE 41) for various shapes of wings? What is the proper value of K (Smeaton Coefficient) in the FIGURE 41 equations? Why did their revolutionary wing warping system fail to control their 1902 glider's roll as expected?

## The Wrights' Dilemma

Between 1900 and 1901, Wilbur and Orville Wright had made several changes to the curvature of their glider airfoils. In 1900 they decreased the airfoil curvature from that used by Otto Lilienthal. In 1901, they increased the curvature back to that of Lilienthal and finally in 1901 they decreased the curvature again. Still, in each of these instances, the wing performed much more poorly than Lilienthal's data had led them to expect.

The Wrights began to suspect the accuracy of Lilienthal's data as well as all other existing data in the field of aerodynamics. They were frustrated. The brothers did not believe they could rely on any existing lift and drag data and yet they knew they had no ability to predict the lift and drag forces on any wing shape

they might want to utilize. They needed to develop that capability.

## Preliminary Tests Devices

The Wrights' first meaningful study of the lift and drag of wing models made use of a modified bicycle, which was ridden to create airflow over test wings. The measuring device was a horizontally oriented bicycle wheel that was free to rotate about a vertical axle mounted in front of the bicycle's handlebars, as shown in FIGURE 42. Wilbur mounted a flat plate having an area of 0.66 square feet on one side of the wheel rim oriented perpendicularly to the wind.

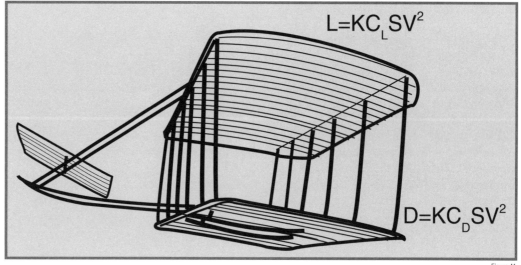

$$L = KC_L SV^2$$

$$D = KC_D SV^2$$

Figure 41

On the other side of rim, he mounted a curved plate having an area of 1 square foot oriented at an angle of 5 degrees relative to the wind, as shown in FIGURE 43. The plate sizes, angles, and locations of the two plates were selected so that both plates would generate equal turning forces on the rim of the wheel; however, each plate would turn the wheel in the opposite direction from the other. If Lilienthal's data were correct, when wind blew on the plates, the forces they transferred to the wheel would be balanced and the wheel would not turn. If the wheel turned in the wind, Lilienthal's data would be considered erroneous. Wilbur said:

*"If I find that (the wheel) really does so (balance and not turn) no question will remain in my mind that these tables are correct."*

Wilbur selected a relatively calm day for his tests. He rode his bicycle along a course that was at right angles to the wind with the wind on one side. After reaching the end of his course, he turned and rode back with the wind on the other side in order to negate any wind effects on the data. In a letter to Chanute on October 6, 1901, Wilbur reported that, when riding at speeds of 12 to 15 miles per hour on the bicycle, the wheel turned! In fact, he had to increase the orientation angle of the curved plate to 18 degrees, more than three times the original angle, before the curved plate would generate enough force to balance the flat plate, as shown in FIGURE 43. Thus, it appeared that Lilienthal's data probably was very much in error.

The bicycle wheel device was not adequate for accurate testing, nor was a subsequent small box-like device with wind blowing through it. However the results from both were the same: Lilienthal's data still appeared to be flawed.

Recognizing that they had devised an excellent concept for aeronautical testing in the small "trough" as Wilbur called it, the brothers decided to build a much larger trough-like device that would enable them to conduct accurate tests on a wide variety of sample wings in a short period of time. With this apparatus, not only could they check Lilienthal's data thoroughly, but they also could develop a catalog of lift and drag data for many different airfoil shapes, which would be useful as a design tool for future aircraft.

## The Wright 1901 Wind Tunnel

Contrary to popular belief, the Wrights did not invent the wind tunnel. In 1871, Francis Wenham of Great Britain, and a colleague, John Browning, had conducted the first tests in which they studied the reaction on small objects mounted in a man-made air stream flowing in a wooden channel.

The Wright's 1901 wind tunnel is shown in FIGURE 44. It was a wooden box, sixteen inches square by six feet long. A fan for generating an air stream was propelled by the shop's gas engine via the overhead shafting and pulleys that normally were

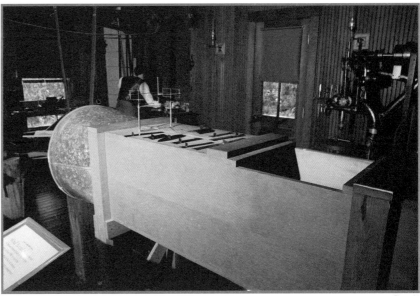
Figure 44

used to operate shop machinery. FIGURE 45 (page 44) is a view of the wind tunnel's air inlet. Note the two-bladed fan and the vanes that reduced the swirling flow from the fan. A tunnel speed from 25 to 35 miles per hour was obtained.

Looking at the tunnel from the orientation of FIGURE 44, the air flowed from left to right, drawn into the tunnel by the fan through a bell-shaped shroud on the left side. After passing through the vanes, the air was straightened further by a sheet-iron, honeycomb grid. The aerodynamic force-measuring instrument, called a balance system, was placed on the tunnel floor near the right end of the tunnel with a test wing model mounted vertically on it. After flowing past the test wing, the wind was exhausted into the room from the left end of the tunnel.

An opening was placed on the top at the left end of the tunnel above the balance system to permit the operator to read the lift or drag measurement directly from a pointer on the balance system. By varying the angle at which the balance system was placed on the tunnel floor, the operator could vary the angle of attack of the wing model.

The proper adjustment of the tunnel was a difficult task.

*"Our greatest trouble was obtaining a perfectly straight current of wind"*, Wilbur told Chanute. Later Wilbur wrote:

FIGURE 44
Wright 1901 wind tunnel.

FIGURE 45
Inlet of the Wright 1901 wind tunnel showing the straightening vanes and fan blades

FIGURE 46
Wright wind tunnel balance used to measuring lift at various angles of attack

FIGURE 47
Wright wind tunnel balance used to measure the ratio of drag: lift at various angles of attack

Figure 45

"We spent nearly a month getting a straight wind."

And straight it was. The tunnel was reputed to maintain flow that varied by only 1/8 of one degree from inlet to outlet, thanks to various straightening devices and careful adjustment by the brothers.

The tunnel was completed by mid October 1901 and serious testing was conducted in November and December of that year. The heart of any wind tunnel is the lift and drag balance system. The device shown in FIGURE 46 was the Wrights' Lift balance. They also made a separate balance, shown in FIGURE 47 for measuring the ratio of drag to lift.

These instruments were truly the work of genius. Although they were made of hack saw blades and bicycle spokes (what else?), the balances were very delicate and precise measuring devices. Their crude appearance was deceptive; actually, they were marvels of simplicity and sophistication. Actually, the balances were computers that solved the lift and drag equations mechanically. The output provided a direct reading of the coefficient of lift, $C_L$, from the lift balance and the direct reading of $C_D/C_L$ (coefficient of drag divided by the coefficient of lift) from the drag balance.

Itemizing the features of their wind tunnel balance system in a letter to Chanute on January 19, 1902, Wilbur wrote:

*"It is our belief that the method and construction employed entirely avoids errors from the following sources: Variation in wind velocity; Variation in temperature and density of the atmosphere; travel of center of pressure and variation in angle of incidence owing to movements of the mounting arms."*

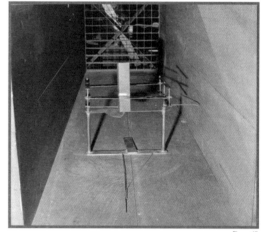
Figure 47

With their wind tunnel and instrumentation built and adjusted, the Wrights were ready to conduct their tests. They finally had the means for obtaining the information required for designing and building a successful flying machine.

## Design Data at Last

Now that they had a wind tunnel, Wilbur and Orville Wright were ready to find the answers to the many aerodynamic questions that had evaded them.

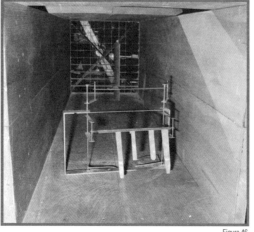
Figure 46

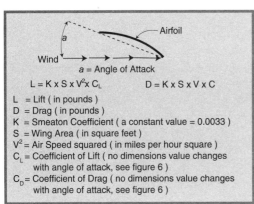

Figure 48

$$L = K \times S \times V^2 \times C_L \qquad D = K \times S \times V \times C$$

L = Lift ( in pounds )
D = Drag ( in pounds )
K = Smeaton Coefficient ( a constant value = 0.0033 )
S = Wing Area ( in square feet )
$V^2$ = Air Speed squared ( in miles per hour square )
$C_L$ = Coefficient of Lift ( no dimensions value changes with angle of attack, see figure 6 )
$C_D$ = Coefficient of Drag ( no dimensions value changes with angle of attack, see figure 6 )

## Screening Tests

The Wrights' began a screening test series in late October 1901 in order to check out the new wind tunnel and acquire a general understanding concerning the effects that variations in an airfoil shape could have on the lift and drag of a wing. They also tested some of their own airfoil ideas.

Their progress was phenomenal. In about one month they tested more than 200 different wing and airfoil combinations. During this time, they adjusted the wind tunnel, determined how to construct wing models quickly, developed a very efficient test procedure and learned which wing design concepts were not worthy of further consideration.

Wilbur kept Octave Chanute well informed of their progress and sent samples of their new data to him. In a letter to Chanute he said:

*"We can make a complete chart of lifts of a surface at angles of attack from 0 degrees to 45 degrees in about an hour."*

Chanute was impressed! He responded: *"It is perfectly marvelous to me how quickly you get results with your testing machine. You are evidently better equipped to test the endless variety of curved surfaces than anybody has ever been."*

And so they were!

## Formal Test Series

The formal tests began on November 22. Wilbur's systematic plan included testing the various wing design variables in such a way that the brothers could learn how each variable affected a wing design. Definitions of these wing design variables are reviewed in Figure 48.

The wind tunnel wing models were made from 20-gauge sheet steel sheet because it was stiff and easy to cut with tin shears. The 20-gauge steel also would provide them with models having a nearly true-to-scale thickness compared to the wings of the full sized glider they planned to build. The brothers found that in only fifteen minutes they could make any wing model in any shape they desired using only tin shears, a hammer and anvil, a file and a soldering iron. If the leading edge of an airfoil needed to be thicker than normal, they would solder an extra piece of material or melt wax or solder on it to fulfill the thickness requirement.

The model airfoils were fastened to the balance systems by mounting clamps soldered to the upper surfaces of the wing models, as shown in Figure 49.

Figure 48
Definition of various wing design variables.

Figure 49
Wright wind tunnel test wing models showing clamps used to fasten the models to the balances.

Figure 49

Figure 50

**FIGURE 50**

Typical wing shapes used in the Wrights' screening tests of more than 200 wings.

**FIGURE 51**

Samples of the lift coefficient vs. angle of attack tabular data obtained during the Wright wind tunnel tests.

Forty-eight models were selected for the formal tests: 43 models were tested for lift and 48 models were tested for drag. Some of the model wings tested are shown in FIGURE 50. The wind tunnel test series included study of the effect of variations in aspect ratio, camber, airfoil shape and leading edge and trailing edge shape. The Wrights compared rectangular wing plan forms with those having tapered tips and also combined the wing models to form biplanes and triplanes. They even varied the gap between the biplane wings to determine the magnitude of the optimum gap between the

wings, since they were planning to build biplanes. And finally, they tested airfoils that had been developed by Otto Lilienthal and Samuel P. Langley, Secretary of the Smithsonian Institution.

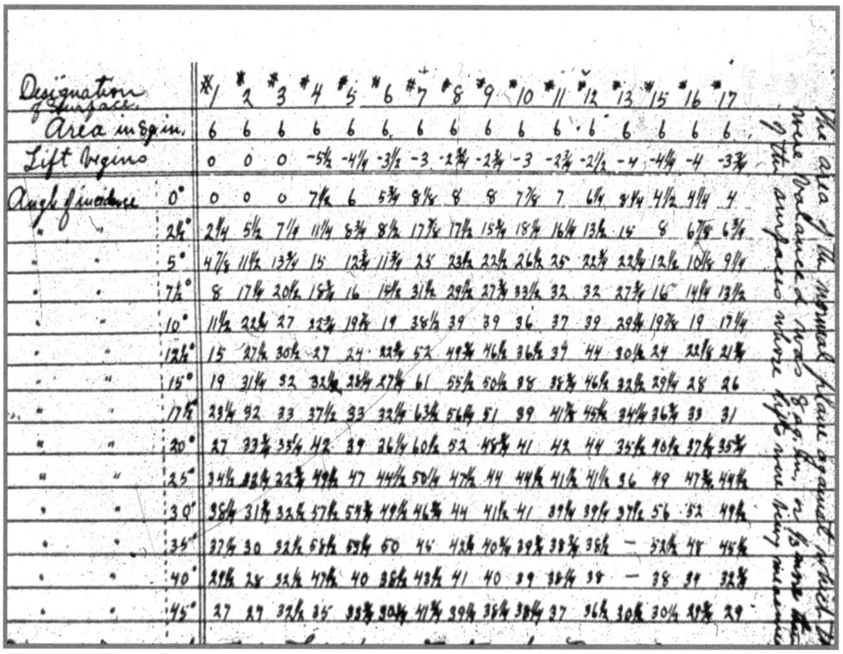

The Wrights were concerned about controlling the airflow into their tunnel. The air from the tunnel was exhausted into the room creating eddies as it whirled around the machinery and workbenches before returning to the tunnel. Wilbur describes the procedure for controlling the airflow:

*"After we began to make our record measurements we allowed no large object in the room to be moved and no one except the observer was allowed to come near the apparatus and he occupied exactly the same position beside the trough (Wilbur's term for the tunnel) at each observation. We had found by previous experience that these precautions were necessary, as very little is required to deflect a current a tenth of a degree, which is enough to very seriously affect the results."*

Referring to the lift and drag equations in FIGURE 41, the data taken during the tests were in the form of lift coefficients, $C_L$, from the lift balance and the drag balance provided data in the form of $C_D/C_L$, the drag coefficient divided by the lift coefficient. These coefficients were measured at fourteen different angles of attack varying from 0° to 45°. Thus for each wing model, a total of 28 lift and drag readings was taken.

The Wrights' data collection method must have been very efficient because Wilbur said he could finish a test in less than an hour. A typical table of their lift data is shown in FIGURE 51, which is a listing of all lift vs. angle of attack data for 17 different wings. This obviously was not the work of "tinkerers"!

Since the brothers had obtained $C_L$ data and $C_D/C_L$ data directly, it was a simple procedure to compute the $C_D$ values by:

$$C_D = C_D/C_L \times C_L ,$$

for each angle of attack.

The tabular lift and drag data was plotted on a graph similar to FIGURE 52, which presents the wind tunnel data for two different airfoil shapes. Note that every

| Designation of surface | #1 | *2 | 3 | *4 | 5 | *6 | #7 | *8 | 9 | *10 | *11 | *12 | 13 | 15 | 16 | 17 |
|---|---|---|---|---|---|---|---|---|---|---|---|---|---|---|---|---|
| Area in sq in. | 6 | 6 | 6 | 6 | 6 | 6 | 6 | 6 | 6 | 6 | 6 | 6 | 6 | 6 | 6 | 6 |
| Lift begins | 0 | 0 | 0 | -5½ | -4½ | -3½ | -3 | -2¾ | -2¾ | -3 | -2¾ | -2½ | -4 | -4½ | -4 | -3¾ |
| Angle of incidence 0° | 0 | 0 | 0 | 0 | 7½ | 6 | 5¾ | 8½ | 8 | 8 | 7⅞ | 7 | 6¼ | 8¼ | 4½ | 4¼ |
| 2½° | 2¼ | 5½ | 7¼ | 11¼ | 8¾ | 8½ | 17¾ | 17½ | 15¾ | 18½ | 16¼ | 13½ | 14 | 8 | 6⅞ | 6¾ |
| 5° | 4⅞ | 11½ | 13¾ | 15 | 12¾ | 11¾ | 25 | 23¾ | 22½ | 26½ | 25 | 22½ | 22½ | 12½ | 10½ | 9¼ |
| 7½° | 8 | 17½ | 20½ | 18¾ | 16 | 14¾ | 31¼ | 29¼ | 27¾ | 33½ | 32 | 32 | 27½ | 16 | 14¼ | 13¼ |
| 10° | 11½ | 22½ | 27 | 22¾ | 19¼ | 19 | 38½ | 39 | 36 | 37 | 39 | 29¼ | 29¼ | 19 | 17½ | |
| 12½° | 15 | 27½ | 30½ | 27 | 24 | 22½ | 52 | 41¾ | 36½ | 39 | 44 | 30½ | 24 | 22½ | 21½ | |
| 15° | 19 | 31¼ | 32 | 32½ | 28½ | 27½ | 61 | 55¾ | 50¼ | 58 | 55½ | 46½ | 32½ | 29½ | 28 | 26 |
| 17½° | 23½ | 32 | 33 | 37½ | 33 | 32½ | 63½ | 56½ | 51 | 59 | 41½ | 45½ | 34¼ | 36½ | 33 | 31 |
| 20° | 27 | 33½ | 35½ | 42 | 39 | 26½ | 60½ | 52 | 45¾ | 41 | 42 | 44 | 34½ | 40½ | 37½ | 35¾ |
| 25° | 34½ | 32½ | 32½ | 44½ | 47 | 44½ | 50½ | 47½ | 44 | 44½ | 41½ | 41½ | 36 | 49 | 47½ | 44½ |
| 30° | 35½ | 31½ | 32½ | 57½ | 59½ | 49½ | 46½ | 44 | 41½ | 41 | 39½ | 39½ | 37 | 56 | 52 | 44½ |
| 35° | 37½ | 30 | 32½ | 65 | 65½ | 60 | 44 | 42½ | 40½ | 39½ | 38½ | 38½ | — | 52½ | 54 | 45½ |
| 40° | 21½ | 28 | 32½ | 47½ | 40 | 36½ | 43½ | 41 | 40 | 39 | 36½ | 38 | — | 38 | 54 | 32½ |
| 45° | 27 | 27 | 32½ | 35 | 33½ | 30½ | 41½ | 39½ | 38½ | 37 | 36½ | 30½ | 30½ | 41½ | 29 | |

Figure 51

airfoil has a unique lift and drag coefficient graph. Thus the lift and drag coefficients vary, not only with angle of attack, but also with airfoil shape.

## Key Number Fourteen • The World's Largest Set Of Wind Tunnel Data

The Wrights learned many lessons from these wind tunnel tests. In particular, they learned that long, narrow wings were more efficient than those that are short and stubby. Also, wings having parabolic shaped airfoils and tapered tips were found to be more efficient.

The formal testing was completed on December 7, 1901, only two weeks after they had begun. Katherine said in a letter to her father:

*"The boys have finished their tables of the action of the wind on various surfaces, or rather they have finished their experiments. As soon as the results are put in tables, they will begin work for the next season's bicycles…"*

The Wrights regretfully decided to discontinue their tests. But their discipline prevailed. They had obtained all the data they had sought. Now they must get back to manufacturing the bicycles needed for the coming year's business, although they would have liked to check out a few more ideas. Wilbur put it best in a letter to Chanute:

*"We saw that any further time consumed by additional testing now would seriously impair our chance of a trip to Kitty Hawk next fall."*

He also told Chanute;

*"Unless I decide to devote myself to something other than a business career (bicycle business), I must give closer attention to my regular work for a while."*

Octave Chanute was shocked they were discontinuing their tests and offered to seek funding from his friend, Andrew Carnegie. However, Wilbur did not want to be beholden to anyone and politely refused, saying:

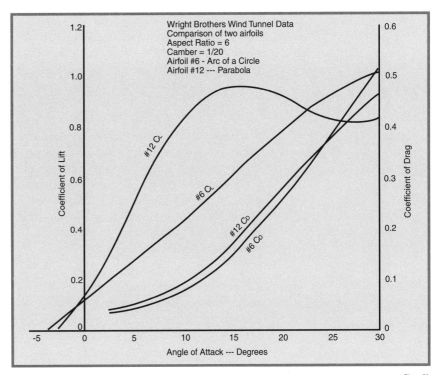

Figure 52

*"I think it possible that Andrew is too hardheaded a Scotchman to become interested in such a visionary pursuit as flying."*

During the fall and early winter of 1901, the brothers' moods had swayed emotionally from agony to ecstasy, sinking to the depths of despair and nearly giving up on their attempt to fly, to their present psychological high resulting from their successful wind tunnel tests. They now realized that they were the absolute world leaders in the field of aviation. They, and only they, possessed a body of knowledge based on accurate tests capable of being used to design successful airplanes. It is clear that the possession of the largest reliable set of aeronautical information in the world was certainly the fourteenth Key to the Wright Brothers' success.

### Smeaton Coefficient Tests

The Wright Brothers now had the necessary values for coefficients of lift and drag to design their wing. But they needed to address their second question: What is the correct value for the Smeaton Coefficient? Its value directly affected the computed values of lift and drag.

When wind blows on a structure, it exerts a dynamic pressure that impacts the

FIGURE 52
Graph of lift coefficient vs. Angle of attack for airfoils #6 and #12. Wilbur Wright indicated that airfoil #12 had the highest dynamic efficiency. This airfoil, slightly modified, was used for the 1902 glider.

structure with a moving force, as opposed to the stationary pressure imposed by a pile of bricks lying on a table. It is this dynamic pressure in hurricanes that destroys trees and buildings. Yet we depend upon lower wind-speed dynamic pressures to enable an airplane to rise from the ground and fly.

Chapter Three described the work of a famous British engineer, John Smeaton in 1759. He essentially asked the question: How strong is the wind? Then he set out to find the answer. After careful experiments using the best equipment that was available at that time, he determined that the wind pressure, P, would be:

$$P = 0.005 \, V^2,$$

where P is the pressure in pounds per square foot and $V^2$ is the wind speed in miles per hour squared. The number 0.005 was known as the Smeaton Coefficient of air pressure.

Aeronautical experimentalists had used Smeaton's coefficient religiously since its introduction in 1759. However, by the beginning of the 20th century a number of scientists, including the Wrights, were beginning to question its validity. Chanute had collected 50 values of K varying 0.0027 to 0.005 from his friends around the world. Still, Chanute believed in the time-honored value of 0.005. To make matters more confusing, European experimentalists had converted the coefficient to metric terms for their calculations, and in rounding off the numbers, had effectively increased the number to 0.0055.

Referring to the lift and drag equations in FIGURE 41, one can see that since the Smeaton Coefficient, K, appears in both equations, the values of both lift and drag depend directly on the Smeaton Coefficient. Thus if K were incorrect by a certain percentage, both lift and drag also would be incorrect by the same percentage,

provided that all of the other numbers were correct.

## KEY NUMBER FIFTEEN • DETERMINING THE CORRECT SMEATON COEFFICIENT

The Wrights had noted during their 1900 and 1901 glider tests that their measured lift and drag forces, both, were only about one-half as large as those predicted by their computations using the time-honored value of the Smeaton Coefficient. Something had to be wrong! The brothers became convinced the Smeaton Coefficient predicted wind pressure that was too high. In letters to Chanute, Wilbur wrote:

*"Prof. Langley and also the Weather Bureau officials found that the correct coefficient of pressure was only about 0.0032 instead of Smeaton's 0.005".*

Later Wilbur wrote:

*"I am firmly convinced that it [0.005] is too high."*

To test their hypothesis, the Wrights conducted a series of ingenious experiments described by Harry Combes, former president of Gates Learjet, in his PBS documentary *"How Strong is the Wind?"* Combes indicated that the Wrights dismantled their 1901 Glider and measured the drag on one wing by attaching a spring scale to two lines extended to the wing. Then Wilbur flew the wing like a kite in a strong wind and measured the drag force as the wing pulled horizontally against the spring-scale, as shown in FIGURE 53. At the same time, the Wrights measured the speed of the wind with the Richards anemometer lent to them by Octave Chanute, as well as the angle of the wing relative to the wind direction.

Now the Wrights had all of the information required for an independent computation of the Smeaton Coefficient. Referring to FIGURE 41, the Drag equation is:

$$D = K \times C_D \times S \times V^2.$$
**Solving for K, we have:**
$$K = \frac{D}{C_D \, S \, V^2}$$

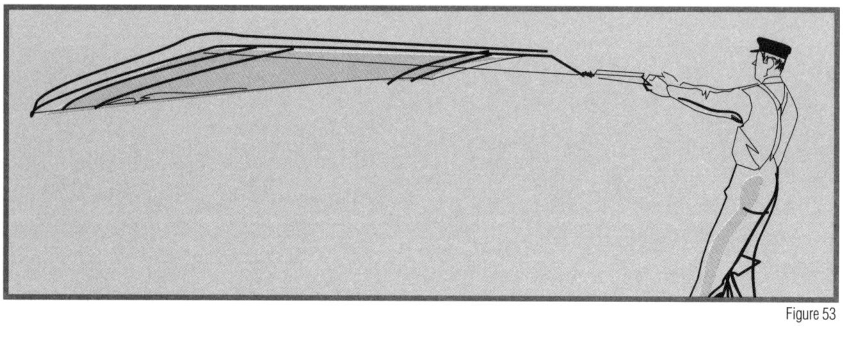

FIGURE 53
 Use of spring scales for measuring the drag on a wing suspended in a strong wind.

Figure 53

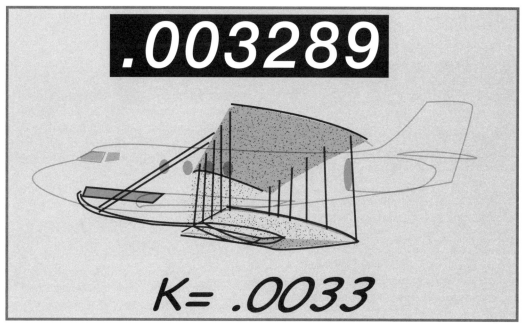

.003289

K= .0033

Figure 54

From their test, the Wrights measured Drag = 10.36 pounds and the airspeed = 20 miles per hour. They knew the area of the one wing was 145 square feet and their wind tunnel tests had indicated that the drag coefficient of their wing was: $C_D = 0.0541$ at the angle at which the wing had been tested. Since the Wrights knew the values of all the terms except K, they substituted the above numbers into the equation for K and obtained:

**K = 0.0033**

How good was this new number? Extremely good! Combes indicated that the sea level value of the Smeaton Coefficient used currently by his aeronautical engineers when designing jet aircraft is 0.003289, as shown in Figure 54. Therefore, the correction of the age-old Smeaton Coefficient to such a precise degree compared to current standards was truly a remarkable achievement.

The Wrights now knew how strong the wind was: $0.0033V^2$!

Now, when the Wrights used their own value for K in computing the lift and drag from their glider kite tests, the test data agreed much more closely with the their theoretical predictions.

This correct determination of the value of the Smeaton Coefficient was the fifteenth Key to the Wright Brothers' success. This

is just one of many instances affirming the Wrights' true use of the engineering approach in designing their gliders and airplanes. They did not just "tinker around in their shop until they came up with an airplane that would fly," the belittling statement that many media writers enjoy using. The term tinker is defined as "an unskillful attempt to mend or improve; to bungle or to botch." Tinkering is a totally inappropriate description of Wilbur and Orville's work and genius.

Now the Wrights' had answered two of the three questions concerning the design of their new glider; they could proceed with confidence. The third question concerning the unexpected behavior of their control system would be addressed later.

Wilbur and Orville's accomplishments were at this time even more astounding when one realizes that, starting with no prior knowledge in aeronautics, they had become self-taught world leaders in aeronautics in only a little more than two and one-half years. At the beginning of 1902, and for many years to come, no one in the world would know as much about the field of aeronautics as these "preacher's kids" who lived in Dayton, Ohio.

FIGURE 54
Comparison of the Wrights' value of K with the sea level value of K currently used to design jet aircraft.

# Chapter 8 – The Glider that Came from a Box

The year 1902 dawned with the Wright brothers full of excitement and high expectations that they were nearly ready to fly. They also were to find out that 1902 probably would be one of their busiest years.

They would to help their father in his political battles with the United Brethren Church; spend endless hours communicating with Octave Chanute on a variety of subjects; build what they hoped would be a revolutionary flying machine; conduct flight tests and acquire flying experience in their 1902 glider at Kitty Hawk and begin development of their 1903 powered machine.

They had good reason for their high hopes. They had solved two of the three problems that challenged them in the fall of 1901. They had determined the correct value of the Smeaton Coefficient of Air Pressure, K, and had completed their wind tunnel experiments. Now the Wrights had a body of reliable data with which they could design a successful aircraft. On the other hand, they had not determined the cause of their 1901 glider's erratic performance during banked turns.

Wilbur recalled his bitter disappointment when he attempted to bank the 1901 glider to the left, for example. In banking left, the glider would tilt its left wing downward and its right wing upward, as expected, and begin a banked turn to the left. However, part way through the turn, the glider would slowly decrease its tendency to turn to the left. Then, without a change in the pilot's control and without changing the tilt of its wings, the glider would start to turn to the right with the left wing still tilted downward. To make matters worse the glider also would start to slide downward to the left as it spiraled to the right. This performance was both frightening and puzzling and the Wrights' had no explanation for it.

With this control uncertainty and in view of the fact that the Wrights had no

assurance that the wind tunnel data were directly applicable to full size aircraft, the brothers decided against building a powered airplane at this juncture. Instead, they planned to build another glider based on their wind tunnel data as well as to add a rear, vertical, non-movable fin as a means of countering the control problem they had faced. In addition, their plan was to make hundreds of glides in order to become experienced aviators.

## KEY NUMBER SIXTEEN • A GLIDER DESIGN OPTIMIZED IN A WIND TUNNEL

The 1902 glider was almost literally "a glider that came from a box"; i.e. a wind tunnel shaped like a box. Work on the glider probably began about May of 1902.

After the wind tunnel tests were completed, the brothers plotted all of the data and analyzed it to determine which relationships of the design variables were optimal for their 1902 glider design. We see a note by Wilbur on the airfoil charts that Airfoil #12 had the highest dynamic efficiency. What did he mean by this?

Not only must an effective airfoil have sufficient lifting capacity, but it also must provide its lift with the least drag penalty in order to reduce the amount of power required to propel the aircraft. So Wilbur must have asked the question:

*"Which airfoil has the highest value of L/D (Lift divided by Drag) in the range of wing angles of attack, (in which we plan to be flying?)"*

FIGURE 55 is a graph comparing L/D of Airfoil #12 with that of Airfoil #6 at angles of attack from minus 5° to plus 30°. With a maximum L/D ratio of 9.5 at an angle of attack of 5°, the parabolic-shaped Airfoil #12 was not only better than Airfoil #6, it had a higher maximum value of L/D than all other airfoils tested. So the Wrights selected a slightly modified version of Airfoil #12 for the 1902 glider wing because it was expected to provide the highest efficiency.

The optimal glider wing design relationships discovered from the wind tunnel tests are shown in FIGURE 56. The

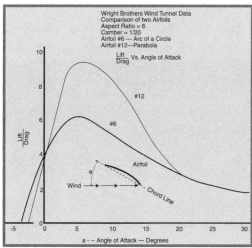

Figure 55

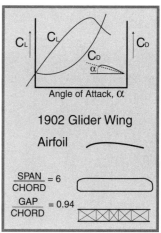

Figure 56

design included: a biplane configuration, a wing having tapered trailing edge at the tips, a span 6 times longer than the chord length, the gap between wings equal to 94 percent of the chord length and a parabolic airfoil having a maximum camber of 1/30 of the chord length. The position of maximum airfoil camber was 1/3 of the chord length back from the leading edge. In addition, the Wrights had the lift and drag coefficients vs. angle of attack for this wing in the graphical form shown on a generic airfoil graph in Figure 56. The Wrights' wind tunnel accomplishments are more astounding when one notes that even today general aviation aircraft follow the Wrights' trend of having wing spans that are about 6 times longer than their chords. The Wrights learned it first!

Now, the Wrights needed to answer the question: how large must the glider wing be made? The brothers had estimated that the weight of the glider alone would be 116 pounds and that the wing would have to lift from 250 to 260 pounds when carrying a pilot.

Based on this weight-carrying requirement and the information in Figure 56, the Wrights developed the final design of the 1902 glider. It had a wingspan of 32 feet, a chord of 5 feet, a gap of about 4.5 feet, and an area of 305 square feet. The wing was about two inches thick. The 1902 glider became the largest glider that had ever been made, even surpassing the Wrights' previous world-record-sized 1901 glider, which had a wing area of 290 square feet.

The addition of the rear, non-movable vertical fin was a major change from their previous gliders. They hoped that this fin would provide a means for overcoming their erratic turning problem. The Wrights reasoned that as the glider started to slide to one side, the vertical fin would act like a weather vane that would keep the glider pointing in the desired direction of a turn and resist the glider's undesirable tendency to reverse turning directions. With this added rear fin, the overall glider length was 14 feet.

The Wrights' design work resulted in the beautiful 1902 glider shown in Figure 57 (page 52), which is a full-sized flying reproduction of the machine that had been on display at Wright State University. This glider was built by a Virginian, Rick Young, and was flown by his daughter at Kitty Hawk for a WGBH television show, "The Wright Stuff". Katie Hudec, a relative of the Wrights, was one of the directors. During this presentation, the glider's smoothly undulating flight set to soaring, graceful music was a synergistic symphony of sight and sound that Wilbur and Orville also must have felt that fall in 1902 when eventually they flew their 1902 glider.

The Wright Brothers' 1902 glider was the first aviation vehicle ever to have its design optimized in a wind tunnel. The ability to conduct this type of wind tunnel design optimization is the sixteenth Key to the Wright Brothers' success.

## 1902 Flight Test Preparations

Construction of the glider probably began in May 1902. On August 20, 1902, Katharine Wright wrote to her father:

*"Will and Orv … are talking of going [to Kitty Hawk] next Monday, though Will thinks he would like to stay and see what happens at Huntington next week .*

*[…This was one of the Bishop's political battles in Indiana in which Will had been actively participating during the year…]*

*They really ought to get away for a while. Will is thin and nervous and so is Orv. They will be all right when they get down in the sand where the salt breezes blow, etc. They insist that, if you aren't well enough to stay out on your trip, you must*

**Figure 55**
Lift : Drag ratio vs. angle of attack data from Wright wind tunnel tests for airfoils # 6 and # 12.

**Figure 56**
Optimized design relationships for 1902 glider obtained from the 1901 wind tunnel tests.

Figure 57

FIGURE 57
Full-sized reproduction of
1902 Wright Glider made
by Rick Young on display
at Wright State University.

*come down with them. They think that life
at Kitty Hawk cures all ills, you know."*

Katherine continued:

*"The flying machine is in process of
making now. Will spins the sewing
machine around by the hour while Orv
squats around marking the places to sew.
There is no place in the house to live but
I'll be lonesome enough by this time next
week and wish that I could have some of
their racket around."*

Wilbur and Orville left by train on their
newest adventure to Kitty Hawk on August
25, 1902, hopeful that their new knowledge
that "came from a box" would enable them
both to learn how to fly at last.

Dan Tate met Wilbur and Orville at the
Kitty Hawk wharf on Aug. 28, 1902. Tate,
the half brother of Kitty Hawk Postmaster
Bill Tate, had agreed to assist the brothers
during their flight tests at Kitty Hawk
during 1902. With baggage loaded on
his spritsail boat, Tate sailed them to the
campsite near the Kill Devil Hills where
they had erected a building in 1901. Almost
three days of pleasant, uneventful travel had
elapsed since they had left Dayton.

The brothers were glad to get back but
were shocked by the condition of their
camp building. The strong winter winds had
damaged part of the foundation and both
ends had settled two feet causing a hump
in the roof. Obviously, much repair work

would have to be done before they could
even begin assembling their new glider.

They had planned to stay in Kitty Hawk
much longer this year. Consequently, they
had decided to make life in camp more
comfortable by erecting an addition to their
building having a kitchen, an eating area
and beds built aloft just below the rafters.
Ten days after they arrived, with Tate's
help, they had repaired the old building,
erected the new addition and had drilled a
16-foot well. FIGURE 58 shows the interior
of their new kitchen with its neat rows of
canned goods arranged by Orville. A tent
was erected to increase storage area for the
glider and parts. Visitors came but were
quickly dispatched. Orville killed two mice
and chased away some hungry razorback
hogs.

Glider assembly began on September 8
and ten days later their 1902 glider stood
ready to test. During assembly, they had
tested each wing as well as the total biplane
wing assembly by flying each assembly as
a kite. The results were encouraging. This
glider promised to be significantly more
efficient than the 1901 glider, fulfilling the
expectations that their wind tunnel data
had foretold.

Along with the usual front rudder
assembly, the brothers added the new
rear vertical fin, which was constructed
of two rectangular side-by-side biplane-
type structures oriented vertically. The
completely assembled glider was first flown
like a kite as in FIGURE 59. Note that the
glider is soaring nearly straight overhead
indicating that it was very efficient (high
lift/drag ratio). The brothers were elated.
Not only was the 1902 glider the most
efficient that they had ever built, but for the
first time their glider's performance agreed
with their theoretical predictions!

## Preliminary Flights

On Sept. 19, Tate and the brothers took
the glider to one of the small hills where
Wilbur made 25 very tentative glides in one
afternoon. FIGURE 60 (page 54), shows the
glider being launched. With its long narrow
wings, the 1902 glider truly was an object of
beautiful simplicity and grace.

As it flew over the sandy landscape, the 1902 glider flew as well as it looked. Both the roll and pitch controls were very responsive and the brothers were convinced that the rear rudder had overcome their 1901 glider's problems. On the first two days, Wilbur made about 50 flights --- about half were on the big hill. A new control-actuating device had been added: a hip cradle for the pilot to lie in so that he could warp the wings by swiveling his hips from side to side. The error-free use of the new hip cradle device was to cause some problems for Wilbur.

Unfortunately, during some of the longer glides, Wilbur had an uneasy feeling. When a wing tip would drop, the glider would tend to slide toward that lowered wing. Lack of familiarity with the new hip-activated control caused problems, too. Once, after Wilbur had glided for about 200 feet, the left wing tip lifted, the wind caught under it and by using the new controls improperly, he caused the nose of the glider to pitch abruptly upward. The result was a stall, a dive to the ground and a hard landing with the lower wing hitting the ground first. To everyone's surprise, Wilbur was not injured and the glider was not damaged!

Orville first flew on Sept. 23. He began flying on the small hill, and then graduated to the big hill for longer flights. He progressed rapidly showing excellent aptitude for flying. His longest glide was 160 feet on his first day. However, Orville still had some lessons to learn. He also stalled like his brother and came down with a hard 'thump!' Again, no injury or damage occurred. A few flights later he wasn't as lucky.

While gliding along smoothly, Orville noticed that the right wing had risen more than he desired. While trying to bring the wing tip down, it mysteriously rose even more and the nose of the plane pitched violently upward to an angle of about 45 degrees. Despite cries of alarm from Wilbur and Tate, the glider stalled, started to move backward and then crashed from a height of about 30 feet. Orville found himself, in;

"…a heap of flying machine, cloth, and sticks in a heap with me in the center without a bruise or a scratch."

Figure 58

Nevertheless, the brothers were jubilant. In only three days, they had made more successful glides than Wilbur had accumulated during the last two years. Also, the glider could be repaired easily. This crash turned out to be the only severe structural damage the 1902 glider ever received during nearly 1000 flights for which it was used. Its resilience was a real testimony to its structural design and durability.

As expected, human visitors began to arrive. On Sept. 30, one of their older brothers, Lorin Wright, was seen trudging over the sand carrying a load of baggage and on Oct. 1, George Spratt, an assistant of Octave Chanute, appeared. They expected Chanute accompanied by another assistant, Augustus Herring, to arrive in a few days.

FIGURE 58
New kitchen area added to the 1901 work shed with beds aloft.

FIGURE 59
1902 Glider flown like a kite.

Figure 59

Figure 60 Courtesy of Special Collections and Archives, Wright State University

FIGURE 60
 1902 Glider in flight right after launch

FIGURE 61
 1902 Glider in flight

The Wrights resumed gliding after four days of downtime for glider repairs. Their flying skills improved rapidly. With Lorin and Spratt's assistance, they made 25 or more flights per day like that shown in FIGURE 61. By Oct. 2, Wilbur had made glides of 550 feet and Orville's longest flight was 216 feet. On occasion, the glider would not respond reliably to the wing warp control. Something still was wrong.

The brothers were pleased with the front rudder control. It had saved both of their lives in severe crashes – crashes similar to those that had killed both Otto Lilienthal and Percy Pilcher, who had gliders with horizontal tails mounted on the rear.

On the other hand, the Wrights began to wonder if the rear fin was helping or hindering. For example, in a left turn any glider or airplane will roll with the right wing raised and the left wing lowered. From this banked position, the 1901 glider, which had no rear fin, would eventually spiral dive to the right. The 1902 machine, which had a rear fin, would perform in the opposite manner. While banked for a left turn, the 1902 machine would first slide to the left toward the lowered left wing and then

would tighten the turn into a spiral dive to the left. This maneuver was the dreaded tailspin that was to claim the lives of many early aviators.

After dinner on the evening of Oct. 2, Wilbur and Orville were desperately analyzing their control problem in an attempt to determine the cause of the glider's dangerous control response. Their discussion evolved into one of the Wrights' famous verbal "scraps", and Lorin and Spratt sat quietly enjoying the healthy arguments. As a result, everyone stayed up later than usual and Orville drank more coffee than was his custom.

Afterward while lying in bed, unable to sleep because of the stimulation of the debate and coffee, Orville ran the theories for the glider's behavior over and over in his mind. Suddenly an idea struck! At last he knew the cause and had a solution for the problem. With a smile, he went to sleep. Tomorrow at breakfast would be soon enough to tell Wilbur.

## Orville's Idea

Orville Wright had prepared breakfast, and his brothers Lorin and Wilbur accompanied by their friend, George Spratt had gathered around the table. Orville told them of his sleepless night, that he knew the cause of the glider's undesirable turning performance and that he had a remedy for the problem.

As his companions listened, Orville reminded them of the glider's flight characteristics in a turn. For example when the glider banked for a left turn, it tended to slide sideways to the left in the direction of the lowest wing tip. While sliding to the left, wind generated by the sidewise movement created a sidewise force on the left side of the fixed rudder, the rudder was pushed to the right and consequently the nose was pivoted even more sharply to the left. This motion resulted in a tailspin.

Orville's solution was to hinge the rear vertical rudder so it could turn either right or left. When the pilot wanted to turn, he would simultaneously activate both the wing warping mechanism and the rudder control in appropriate directions. For example, to turn to the left, the

Figure 61 Courtesy of Special Collections and Archives, Wright State University

simultaneous activation of the wing warping mechanism and the rudder control would cause the left wing to drop into a left bank (left wing tip lowest) and the rudder would turn to the left in order to keep the glider following a smooth banked turn that would prevent it from slipping sideways.

As Orville presented his solution, he winked at his brother, Lorin, expecting Wilbur to say that he had thought of it before or that he had discarded the idea. To Orville's surprise, Wilbur listened attentively, and pondered the thought for a moment. Wilbur said it was a good idea and then offered a refinement. He pointed out that the pilot was already very busy controlling the machine by using his left hand to control the front rudder, his hips to control the wing warping mechanism, and his right hand just to hold on so he would not be thrown from the glider. The Wrights did not have seat belts or other safety restraints for many years. Wilbur's idea was to connect the rudder control to the wing warping control so that when the hip cradle was activated it would simultaneously cause the wings to bank and the rudder to turn in the direction of the lowered wing. This dual control would provide the desired banked turn.

Construction of a new, single-surface hinged rudder began immediately and the controls were changed as required. The new system was ready by October 6. Instead of a fixed, double, rear rudder, they now had a movable, single rear rudder that was half the size of the original.

## Key Number Seventeen • Three Axis Control

The modified glider first was taken to the little hill for a few familiarization glides. Then they all went to the big hill where Lorin took a dozen pictures of Wilbur gliding. The machine flew beautifully. On two of the glides, Wilbur brought the glider to a stop mid-air, pirouetted to the left and landed in a 90° crosswind with no problem. This first successful turn, shown in Figure 62, was made on October 10, 1902. Wilbur continued making complex glides having no trouble with the control system. On one he flew 280 feet making turns at

Figure 62 Courtesy of Special Collections and Archives, Wright State University

large crosswind angles first on one side and then on the other. The glider followed his commands perfectly.

Now it was Orville's turn. Again they took the glider to the small hill to enable Orville to practice while making short glides. Initially he had trouble because he was paying close attention to the wing-rudder control at the expense of neglecting the glider's response to the front rudder pitch control. As a result he caused the machine to bounce up and down considerably. But finally, he also became the master of the craft and was making spectacular flights.

October 10, 1902 was a red-letter day. On this day the Wright Brothers achieved three-axis controlled flight of an aeronautical vehicle for the first time in history. They had successfully controlled their aircraft in Roll, Pitch and Yaw defined in Figure 63, Figure 64, and Figure 65, respectively.

This was a most remarkable invention; particularly since the Wright control system is essentially the same as that used on airplanes today. Instead of warping the wings, we use ailerons, which are hinged controls on the trailing edges of the wing tips. The left and right ailerons deflect in opposite directions, just like the wing warping control of the Wrights. Today, yaw control is achieved by use of a hinged vertical rudder surface in the rear, just like theirs. And like the Wrights, today we operate the aileron and rudder controls simultaneously in a coordinated manner so that the aircraft will maintain a smooth turn without slipping downward to the side or skidding toward the outside of a turn.

FIGURE 62
1902 Glider making the first successful coordinated turn of any heavier-than-air flying machine, October 10, 1902

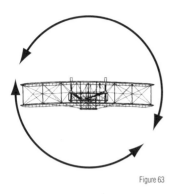

Figure 63

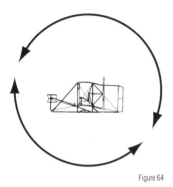

Figure 64

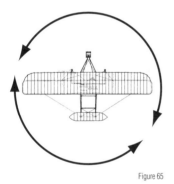

Figure 65

FIGURE 63
Control about the roll axis

FIGURE 64
Control about the pitch axis

FIGURE 65
Control about the yaw axis

FIGURE 66
Three-axis control of the 1902 glider on October 24, 1902. One can see the pilot has actuated the controls to raise the right wing by warping the right wing downward and turning the rudder to the left for a left turn.

Today we also use movable elevators on either a rear-mounted or a front-mounted horizontal control surface (similar to that used by the Wrights) to prevent loss of altitude during a turn. This altitude loss occurs because, when the wing is banked, the lift force is tilted to the side and upward instead of straight up as it is in level flight. Thus in a turn, only part of the lift is pulling upward and there is not enough wing lift to keep the aircraft from losing altitude without help from the elevator of canard.

Their 3-axis control system was the seventeenth Key to the Wright Brothers Success because this mode of control is the essence of flight itself and is the technological advancement that truly elevates the Wrights from all previous and early 20th century aviation experimentalists. Without a doubt, the Wrights' 3-axis control system is the preeminent technological reason for their success. Without 3-axis control, successful flight would not have been possible.

## Learning to be Pilots

The Wrights made a giant step toward learning to fly in 1902. They made flight after flight and became able to control their craft in wind speeds as high as 35 miles per hour in all sorts of orientations, like the pilot in FIGURE 71 using the wing warping and rudder control mechanisms simultaneously to raise his lowered right wing tip and turn the glider to the left. The brothers' flights became longer and longer until Wilbur made a flight of 622 feet and Orville made a flight of 615 feet. One can easily imagine the brothers' excitement as they were launched on flights from the Big Hill. They now were budding aviators.

They also became very efficient in conducting their flights. On one day they made more than 100 glides and during the last six days they made 375 flights. In 1902 they made nearly 1000 glides at Kitty Hawk, which were equivalent to about four hours of flight time. Still Wilbur said:

"…we consider ourselves little more than novices in management [piloting]."

However, these "novices" were at that time the best human flyers that had ever lived on earth.

One has to marvel at the physical expense that this large number of glides per day cost Tate and the brothers. Two men had to lift the 260-pound gilder and pilot and, while carrying it, run down hill in the loose sand until there was enough lift to release it. Then after each flight they had to lug the cumbersome machine uphill again on loose sand. On days when the wind was light, the process was slower and more exhausting: they had to run farther and faster to launch the 260 pound machine and they had no help from the wind while bringing its cumbersome 112- pound airframe back up hill. On very windy days much less exertion was required because they did not have to run far while launching the glider. Also, a strong wind blowing on that large wing surface provided an uphill boost during retrieval. Nevertheless, it was hard work. Fortunately, they were sustained by the exhilaration of flying at last.

Figure 66 Courtesy of Special Collections and Archives, Wright State University

# Chapter 9 – A Machine Designed for Powered Flight

During the waning days of October 1902, the Wright Brothers were formulating the design of their first engine-powered machine: the 1903 Flyer, sometimes called the Kitty Hawk Flyer.

They initially estimated that the Flyer with pilot would weigh 625 pounds, which included 200 pounds for the engine and propeller system. Using this weight estimate and the lift and drag equations, they were able to estimate the wing size as well as the size of the other components necessary to carry the weight. Thus, before they arrived home in Dayton they had developed a preliminary design of their new powered machine.

## 1903 Flyer Design Concepts

Contrary to one of the "darlings of the media", the Wrights did not mount an engine on one of their gliders! Being much larger and heavier and having an engine, the 1903 Flyer had to be a completely new design, although it did utilize many concepts proven by their gliders.

The new Flyer would be a biplane having a wingspan that was six times longer than the chord with tips tapered from the trailing edge forward, a configuration that had been validated by their wind tunnel tests.

The pilot lay prone on the left side of center, and his weight was partially balanced by the engine mounted on the right side of center. Since the engine weighed about 34 pounds more than the pilot, the right wing was made four inches longer than the left wing. More lift on the right side compensated for the weight difference and insured that the Flyer would be balanced from side to side. The control system was like that of the 1902 glider with the pilot using a hip cradle to activate both the wing warping and rear rudder controls simultaneously. As before, the pilot controlled the front rudder (elevator) by a lever in his left hand.

The power plant design demonstrated ingenuity. When any propeller turns, e.g.

Figure 67

to the right an equal and opposite torque (torsion) is exerted on the airplane itself that twists it to the left and causes the plane to bank to the left.

Consequently, the Flyer pilot would have to nullify this unwanted banking tendency by continually putting pressure on the left wing warping control. To avoid this problem the brothers cleverly chose to utilize two counter-rotating propellers; i.e. when facing the rear of the Flyer, the propeller on the viewer's right rotated clockwise and the propeller on the left rotated counterclockwise. The net result would be no torque on the Flyer. Another advantage of using two propellers was that two propellers react with a larger volume of air than a single propeller, and consequently, a lower propeller rotational speed could be used. The propellers were mounted in the rear to prevent any effect on the wing of the turbulent airflow from front-mounted propellers.

As one would expect, the brothers' bicycle experience came to the fore. They selected a chain drive system to transfer the engine power to the two propellers as shown in Figure 67. The longer chain was twisted in order to turn the propeller to which it was attached in the direction opposite to that of the other propeller. Guides, which covered the chains, helped to prevent the chain from jumping off the sprockets and protected the pilot from flying chain in the event a chain failure. Again, we'll fracture another myth!

FIGURE 67
Rear view schematic of chain drive and propeller system.

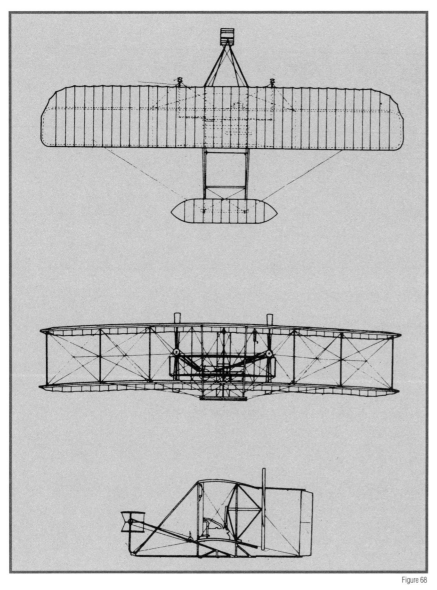

Figure 68

FIGURE 68
Three-view drawing of
1903 Wright flyer.

ground, which provided a higher, safer platform for the pilot during landing as well as additional ground clearance for the Flyer's 8.5-foot diameter propellers.

But how would they take off? Wheels would not be viable in the sand. The answer was the "junction railroad", as Orville called it. The "railroad" was a monorail track 60 feet long made of five sections of 2 x 4 lumber turned on edge and capped with a 1/8-inch thick strip of steel. Ball bearing wheels rolling on the track made a wheeled takeoff possible. The front end of the Flyer was supported on the track by a modified bicycle wheel hub attached to the front cross member of the Flyer's skid. A dolly having two fore-and-and-aft wheels supported the rear portion of the glider. The dolly was wider than the landing skids, so it provided ample support. All wheels were flanged to keep the Flyer from "jumping the track".

During takeoff, one brother would support the wing tip and run alongside the Flyer until it had sufficient speed for the wing warping control to balance the machine. Then the "runner" would release the wing and the Flyer would be free to gain further speed and fly. As the Flyer left the ground, the dolly would be left behind.

## KEY NUMBER EIGHTEEN • AERONAUTICALLY ENGINEERED DESIGN

The Flyer's overall weight estimates became more and more precise as detailed design was completed. The end result was an airplane having a 40-foot, 4-inch wingspan and a wing area of 510 square feet. The wing bracing wires were adjusted so that the wing tips were ten inches lower than the center of the wing. This adjustment, which reduced the Flyer's sensitivity to side gusts, was the result of Wilbur's observations at Kitty Hawk that sea gulls with downward sloping wings were less sensitive to side gusts than other birds having upward slanting wings. The three-view drawing of this airplane is presented in FIGURE 68.

Since the Flyer was so much larger and heavier than the Wright gliders, the wing

The Wrights did not use bicycle chain to transfer engine power to the propellers. Normally a bicycle chain has links that are 1/2-inch-long, whereas the Flyer used heavy, industrial chain having links that were twice as long as well as wider and thicker than bicycle chain links. Although the Wrights did not use bicycle chain, they did use bicycle chain technology.

The Wrights realized that they needed a means for taking off under engine power alone as well as a better means for landing than that provided by their miniscule glider skids. A strong, sled-type runner was designed for landing in the sand. The runners elevated the leading edge of the Flyer's wing about 20 inches above the

| Span: | 40 ft. 4 in. |
|---|---|
| Chord: | 6 ft. 6 in. |
| Gap: | 6 ft. 2 in. |
| Camber: | 1/20 of Chord Length |
| Wing Area: | 510 sq. ft. |
| Length: | 21 ft. 1 in. |
| Weight: | 605 lb. without pilot |
| Orville's Weight: | 145 lb |
| Wilbur's weight: | 140 lb |
| Engine Power: | 12 hp. at 1020 rpm |
| Horizontal Rudder Area: | 48 sq. ft. |
| Vertical Rudder Area: | 21 sq. ft. |

ribs, the structural members that give the wing curvature, had to be both thicker and stronger than the single thin strips of bent wood used for the gliders. In contrast, the 1903 Flyer wing ribs were 1-3/4 inches thick at the leading edge and tapered to 5/8-inch thickness at the trailing edge. To save weight, the ribs were constructed with ash top and bottom cap strips separated by blocks of spruce as shown in FIGURE 69.

Although only the top surface of the glider wing ribs was covered with fabric, both the top and bottom surfaces of the Flyer wing were covered. The brothers used "Pride of the West", a finely woven, unbleached muslin fabric having a thread count of 108 by 108 threads per inch to cover their Flyer wing and tail structures.

How did the Wrights assure themselves that the various parts of the glider were as lightweight as possible and yet strong enough to assure flight safety? Did they guess, build and test, or use some other means? In 1987, Dr. Howard Wolko, a stress analysis authority with the National Air and Space Museum, concluded that the Wrights had indeed used contemporary

1903 stress analysis methods to determine the sizes of all structural members. He said:

*"The 1903 Flyer was the product of a careful aeronautical engineering design for which the Wright's used wind tunnel data, aerodynamic equations, flight test data, propeller design, engine design, material properties data and strength of materials computations."*

Thus, the brothers are recognized as being the world's first true aeronautical engineers. Their use of these principles in designing the 1903 Flyer was the eighteenth Key to the Wright Brothers' Success.

With the airframe designed, the next tasks facing the brothers were perceived to be straightforward: purchase an engine and use marine engineering handbooks to design their propellers. These efforts were to be much more difficult than they expected!

## KEY NUMBER NINETEEN • DESIGN AND CONSTRUCT A SUCCESSFUL AIRCRAFT ENGINE

Wilbur Wright had a mission: buy a gasoline engine suitable for powering the new Flyer. Accordingly in late 1902, letters were sent to a large number of gasoline engine manufacturers asking them to bid on his engine. The engine was required to generate from eight to nine horsepower, to weigh not more than 180 pounds and to operate with very little vibration in order to prevent the airframe from being shaken apart.

By December 3, ten responses had been received, but none could meet the requirements. Responders either were unable to meet specifications, were not willing to take the risk or they did not appear to be reliable. Now, the brothers were faced with a real problem! The so-called straightforward task of buying an engine had turned into one of developing a completely new engine themselves from scratch. Wilbur was especially concerned because he and Orville had plenty to do already and also they had only a little experience in engine design or fine machining of metal parts.

FIGURE 69
1903 wing rib design

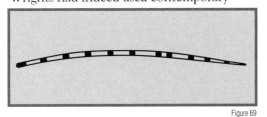

Figure 69

## Charles Edward Taylor's Engine

In June of 1901, the Wrights had hired Charles Edward Taylor, to help them with bicycle repairs and running the shop while they were at Kitty Hawk, North Carolina. Taylor had proven himself to be a trusted employee and an excellent machinist. The brothers and Taylor discussed the problem of building an engine, and Taylor said he could do it. He had worked on automobile engines and the brothers had built their own one cylinder engine for their shop. Taylor was 33 years old at the time, FIGURE 69.

The task was formidable and the Wright's machine tool selection was very limited. They had only a 14-inch lathe, a drill press and a powered bench grinder. That was all! They couldn't afford a milling machine. Later they did buy a band saw. In order to control costs, the brothers wanted to fabricate as many parts in their shop as possible, buy ready-made parts and minimize the number of fabricated parts made outside the shop. Therefore, the engine design also was influenced by the machine tools available in the bicycle shop.

The brothers and Taylor decided to build a 4-cylinder, in-line engine that had the cylinders mounted horizontally instead of vertically as they were in automobiles. They

wanted the engine to lay flat to reduce drag and possibly to spread the engine's weight over a larger mounting area. Taylor also favored the horizontal position because he believed that the Wrights 14-inch lathe would not be capable of making a block having vertical cylinders.

## The First Engine

No engine drawings were made. Orville, working with Taylor, would sketch a part on a piece of scratch paper and Charlie would impale it on a nail over his bench and machine it. Detailed dimensioning and fit tolerances were Taylor's responsibility. The weight of each completed part was recorded in order to maintain control of engine weight.

They decided to use an aluminum engine block to save weight. Fine-grained, gray cast iron was used for the cylinders and pistons. Wilbur tested the cast iron material and found it had excellent strength. The brothers made the wooden patterns for the molds. The engine block, pistons and cylinders were cast in local foundries.

An operating model of the engine, consisting of a single cylinder in its housing, was made first to check their overall design concepts. Charlie said they smeared grease on the model, hooked it up to their shop engine and ran it for short periods while watching its operation to make sure that all parts were running freely. The results were good so they proceeded with the design and fabrication of their engine.

Taylor used very ingenious methods for making the engine parts. For example, the crankshaft, which had four cranks and five bearing journals, was made from a single, 100-pound block of tool steel 31 inches long, 6 inches wide, and 1-5/8 inches thick. First, he traced the shape of the crankshaft on the steel block. The unwanted material was removed by drilling overlapping holes around the perimeter of the part, as shown in FIGURE 71.

Finally, he mounted the part in the lathe and turned the round and flat surfaces smoothly and precisely. The finished part attached to the flywheel is shown in FIGURE 72. Drilling all of these holes in tough, 1-5/8-inch-thick tool steel was a very laborious

FIGURE 70
Charles Taylor, father of the Wright airplane engines

Figure 70

process. A present-day machinist, who built an authentic 1903 Wright engine using only the tools that Taylor used, said that it took three weeks to machine this part alone. Because of this difficulty in machining, historians have noted that a number of early experimental engines made by the Wrights do not have crankshafts in them. One concludes that the Wrights must have used crankshafts from a previous engine in the next engine produced to reduce the time and cost of manufacturing.

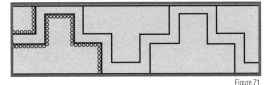

Figure 71

The engine was completed in only six weeks and was tested for the first time on February 12. The test was successful and all were pleased that it ran so smoothly. The next day was Friday the 13th and disaster struck as one might expect. While the engine was running, gasoline dripped on the bearings and the lubricating oil was washed out of the moving parts. Consequently, the crankshaft froze in the bearings and the engine casting blew apart. Bishop Wright wrote in his diary that day"…The boys broke their little gas motor in the afternoon." They salvaged all of the parts they could and arranged to purchase another aluminum casting. On April 20, the Bishop gave another progress report: "…The boys received their aluminum casting for their light-weight engine today."

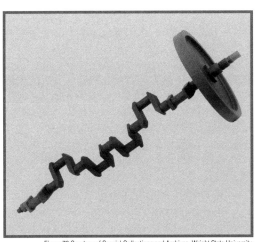

## The Second Engine

Again, Taylor went to work and had the engine operating again in May. By June 28, the engine had been thoroughly tested and the "Three Musketeers" were overjoyed with the results. The engine initially produced 16 horsepower, but because of expansion of parts as the engine heated, the power would drop to a relatively constant 12 horsepower running at a speed of about 1090 revolutions per minute. At 12 horsepower the engine was providing 50 percent more power than the Wrights' estimated requirement for flight. Further, with the engine weighing only 171 pounds and the entire power system weighing about 200 pounds, their power plant weight was within their design specifications. Thus, they would be able to add more weight to any of the Flyers' other parts that would benefit from strengthening.

The engine was very simple. It had no carburetor. Gasoline was gravity fed from a fuel tank mounted high near the upper wing into a can on top of the engine. There, air was mixed with the gasoline and the mixture was vaporized as it passed over hot crankcase surfaces. The vaporized mixture was directed to the manifold and thence to the cylinders when intake valves opened.

There were no spark plugs. Platinum breaker points created a spark when the points opened inside each cylinder's combustion chamber. The energy for the spark was generated by an on-board magneto, which was driven by the flywheel. The initial spark for starting the engine came from an electric coil powered by dry cell batteries, which were not carried during flight.

There also was no water pump. Instead, the cooling water circulated by gravity and by convection through the 5-ft-long radiator mounted vertically on the center front strut of the wing. A splash system was used to provided lubrication for the cylinders.

The engine had only two speeds: top speed and off. Retarding the spark could make minor speed adjustment but this adjustment could be made only while the Flyer was on the ground.

FIGURE 71
How the crankshaft of a Wright engine was rough-cut from a block of tool steel

FIGURE 72
Completed crankshaft attached to the flywheel

Final tests and engine break-in were made using a makeshift 5-foot diameter board for a fan. Because of the shop's small size, the complete power train and propellers could not be assembled or tested there. Assembly and testing would have to wait until the brothers were at Kill Devil Hills.

The successful completion of the 1903 Wright Flyer engine was an amazing feat and the Wrights' acknowledged that the engine was a tribute to Charlie Taylor's skill. Despite their lack of experience and shortage of machine tools, the engine was completed in less that two months, it operated successfully on the first try, its weight was below requirements and its power was 50 percent greater than required.

Further, Taylor believed this was the first 4-cylinder aluminum block engine ever made. Since it later provided the necessary power to assure successful flight at Kitty Hawk, the 1903 engine was the nineteenth Key to the Wright Brothers' success. The completed 1903 Flyer engine is shown in FIGURE 73.

## KEY NUMBER TWENTY • DESIGN OF THE FIRST TRUE AIRCRAFT PROPELLER

FIGURE 73
Completed 1903 Engine

The development of an effective propulsion system was the last major obstacle that Wilbur and Orville Wright had to surmount in their development of their powered flying machine. Wilbur and Orville Wright astutely realized that the shortage of power from their tiny 12-horsepower engine necessitated the development of a highly efficient means for converting the engine's power into sufficient thrust for flight.

The Wright propeller development began in parallel with their engine development in mid-December 1902 shortly after their return from their 1902 triumphs in flight control and piloting at Kitty Hawk. Utilizing the small one-cylinder engine that powered their shop machinery, they conducted a "fan-screw" experiment during which they mounted a 28-inch diameter fan blade on their engine and ran it to determine generally how an aerodynamic propeller worked.

## A Rotating Wing

Initially, the brothers had thought that the propeller development would be a minor task. In their article in the September 1908 edition of Century Magazine, they said:

*"Our [wind tunnel lift and drag] tables made the designing of the wings an easy matter; and as screw propellers are simply wings traveling in a spiral course, we anticipated no trouble from this source. We had thought of getting the theory of the screw propeller from the marine engineers, and then, by applying our tables of air pressures to their formulas, of designing air propellers suitable for our purpose".*

Their literature search was fruitless. No propeller design theory had ever been developed in the marine engineering field. All the brothers found were empirical relationships to use in an expensive, time-consuming ritual of cut and try. The Wrights could not afford the time and cost of such a method. Instead, they decided to develop their own theory that would enable them to design their propeller by mathematical calculation alone.

The more they studied, the more they realized the operation of a propeller was a highly complex process. In a letter to his friend, George Spratt, Wilbur wrote:

*"What at first seemed a simple problem became more complex the longer we studied it. With the machine moving forward, the air flying backward, the propellers turning sidewise, and nothing standing still, it seemed impossible to find a starting point from which to trace the various simultaneous reactions. Contemplation of it was confusing."*

## Scientific Argument

So complex was their study that they filled five notebooks of computations, observations and conclusions.

But the Wrights did solve the problem magnificently using their unique method: vigorous argument. They would sit in their chairs thinking and then one would propose an idea. After quiet contemplation the other would say: "tis'nt". The other would think a while and then say: "tis". Then things would escalate with each giving opinions, drawing sketches, and waving his arms around. Finally the arguments became so heated that at times Katharine had to throw them out of the house.

Charlie Taylor gave us a first-hand account of these "discussions."

*"Both the boys had tempers, but no matter how angry they ever got, I never heard them use a profane word. The boys were working out a lot of theory in those days, and occasionally they would get into terrific arguments. They'd shout at each other something terrible. I don't think they really got mad, but they sure got awfully hot."*

Charlie continued:

*"One morning following the worst argument I ever heard, Orv came in and said he guessed he'd been wrong and they ought to do it Will's way. A few minutes later Will came in and said he'd been thinking it over and perhaps Orv was right. First thing I knew they were arguing the thing all over again, only this time they had switched [sides]. When they were through, they knew where they were and could go ahead with the job."*

The important aspect of these arguments was that they were not contests to see who was right. The aim was always to find out what was right. There was no desire to compromise. They had only the shared objective to develop a scientifically honest and truthful end result.

## A New Propeller Design Theory

The brothers completed their propeller design theory and had made a prototype propeller by mid February; at the same time they had finished their engine. Unfortunately, since the engine flew apart during its second test on February 13, they had to use the shop engine again for their second propeller test. This time they tested a full sized propeller 8.5 feet in diameter. The width of the propeller surface was curved in the shape of a circular arc just like the airfoil of wing #9 from their wind tunnel tests. The wing #9 was selected because it provided the most efficient lift/drag ratio for propeller design of all propellers tested in the angle of attack region in which the propeller would operate. Results of this second propeller test gave the brothers added encouragement about the accuracy of their theory so they proceeded to manufacture the actual propellers for their 1903 Flyer.

Instead of designing flat-bladed paddle-type propellers having a constant blade angle from center to tip, the Wrights designed curved blades having a helical twist, which began with a steep blade angle at the center of the propeller and gradually twisted to a flatter angle at the tip. Since the incoming air strikes a propeller blade at a steep angle at the center and gradually changes to a shallower angle at the tip, the blade twist enabled the incoming air to strike the blade at the angle having the most efficient lift/drag ratio at every point along the blade's radius.

The propellers were made from three, first-quality spruce planks 8-1/2 feet long by 4-1/2 inches wide by 1-1/4 inches thick fanned out about a center hole and glued together as shown in Figure 74. Then, by use of a hatchet, a drawknife, a spoke shave and a chisel the blades were shaped to the required contour. The finished blades

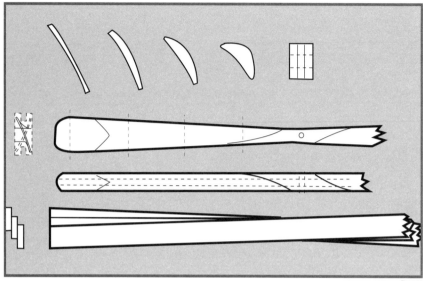

Figure 74

FIGURE 74

Method of laminating 1903 Wright propellers prior to carving and finished contours of Wright 1903 propellers.

were concave on the backside and convex shaped on the front surface. The fanned layout of the three planks enabled the twist to be formed in the blade as it was carved. A canvas lamination was glued to both sides of the tips to reduce the tendency of the blades to split if the tips struck the ground or another object.

As mentioned previously, Wilbur and Orville had elected to use two, counter-rotating propellers in order to obtain a reaction against a greater quantity of air than if they had used only one propeller. Hence, the propellers had to be carved so that one was twisted to the right and the other was twisted to the left. The difference in the direction of twist was achieved by fanning the boards in opposite directions at the beginning step in the manufacturing process.

The propellers performed considerably better than expected. The Wrights' calculations indicated that they expected 90 pounds of thrust from propellers at an efficiency of 66%. Instead, their tests showed that their propulsion system provided 132 pounds of thrust. The reason for the Wright's better-than-expected propeller performance was discovered recently. Ken Hyde, of The Wright Experience, Warrenton, Virginia, conducted tests on Wright Propellers in the NASA Langley Wind Tunnel in the 2002 to 2003 time period. The test results indicated that the 1903, 1904, and the

1911 propellers had efficiencies of 81, 83, and 87 percent, respectively. Hyde has been conducting an in-depth study of the Wright's engineering work, and has built an exact flying reproduction of the 1903 Wright Flyer, including an authentic Wright engine, that will make the official Centennial flight at Kill Devil Hills on December 17, 2003.

The Wright's propeller design placed them far ahead of their competitors. As late as 1908, other American and European aircraft builders used from three to five times more engine power than the Wrights and yet generated less propeller thrust than that developed by the Wright propellers five years earlier.

Since we use essentially the Wrights' theory to design propellers today, their theory is another legacy from them that we enjoy. The higher than expected thrust of the Wright propellers played an important part in assuring their successful flights in 1903. Therefore, we believe that the development of the propeller design theory as well as the superb quality of the propellers designed for the 1903 Wright Flyer constituted the twentieth Key to the Wright Brothers' success.

Charlie Taylor put it best in describing the Wrights' propeller development:

*"I think the hardest job Will and Orv had was with the propellers. I don't believe they ever were given enough credit for that development. They had read up on all that was published about boat propellers, but they couldn't find any formula for what they needed. So they had to develop their own."*

And "their own" is still competitive with our best efforts today

# Chapter 10 – Preparations for Powered Flight

The Wright Brothers realized that they would need a larger hangar building to assemble, house and test their "whopper flying machine", as Wilbur called it, and to provide working space in inclement weather.

Two shipments of supplies, tools and parts were shipped to Kitty Hawk by September 18, 1903, and the brothers embarked on what would be the journey of their lives at 8:55 a.m. on September 23. They arrived a little after noon two days later, pleased to see that their tools, provisions and lumber for the new building were awaiting them. The parts for their machine were to be received later.

## Rebuilding at Kill Devil Hills

Wilbur and Orville were shocked when they saw the severity of damage that their living quarters had received from the preceding winter's storms. Ninety mile per hour winds had moved the Wright's 1902 living quarters two feet from its foundation and had sunk one end one foot into the sand. There were gaping holes in the sides and sand everywhere. They were greatly relieved when they found that the 1902 glider stored in the building was undamaged. Because the building was not insulated, they had fondly called this building "the summer house".

The Wrights also planned to sharpen their piloting skills by flying their 1902 glider when weather permitted. Otherwise, they would work on the new building and on assembly of their new machine. Dan Tate was hired to assist in the carpentry and glider launching.

The new hangar was 44 feet long, 16 feet wide, and 9 feet high. They called this building "the hand car", a take-off on the word hangar, which the French called the buildings, which housed their airships. The new building had doors in each end, hinged at the top in order to provide a maximum opening, as shown in Figure 75.

Figure 75 Courtesy of Special Collections and Archives, Wright State University

## Sharpening Flying Skills

The brothers' first opportunity to fly came on Monday September 28. Tate helped drag the 1902 glider to the Big Hill. About 75 glides were made, some in winds of 31 miles per hour. It was exhilarating. Now they were learning to soar over the slopes, electing to stay aloft while staying motionless over the ground. They made from 12 to 15 flights of more than 20 seconds duration, one of which had a 26 second duration. This type of flying gave them practice of maintaining equilibrium when the glider was buffeted about by the wind.

Wilbur and Orville made about 300 flights, improving their skills all the time. On October 26 they made their best flights, which again broke the world gliding records that they seemed to set daily. In breezes of 13 to 18 mph, they soared nearly motionless for durations varying from one minute, five seconds to one minute, 11 seconds. Figure 76 shows the highest flight they made, with their two buildings below and distant.

Figure 75
1903 flyer in front of new hangar on the left and the living quarters on the right.

Figure 76
High flight above the camp site in the 1902 glider during 1903 practice flying sessions.

Figure 76 Courtesy of Special Collections and Archives, Wright State University

FIGURE 77
Langley's Aerodrome on houseboat ready to launch

## Battling the Elements

October 8 marked a major change in events. The last shipment containing parts for the new machine arrived in conjunction with the completion of the new hangar. The hangar was completed none too soon because dark storm clouds began to appear. The ensuing storm descended upon them with unprecedented fury that prevented them from sleeping that night. The next day they found water had been driven into their living quarters.

As the storm raged, a corner of their roof suddenly began to give way. The entire roof was in danger of being lost momentarily. So in cyclonic wind and driving rain, Orville and Wilbur erected a ladder. After two tries and with much difficulty the roof was nailed in place. Later Orville told Wilbur that the wind was blowing so hard that it misdirected his hammer blows so that "…three licks out of four hit the roof or my thumb instead of the nail."

The storm continued four days. Later Dan Tate told them that the storm had broken all records for persistence and had been equaled by few in velocity. Five vessels had been driven ashore in the area.

This storm was only the beginning of the tortuous weather that they were to endure. Many days were so cold that they could not work. Some morning's water in their

washbasins had frozen solid to the bottom. They referred to nights in terms of the bedding they needed to keep warm; i.e. 5-blanket, two quilt, fire, and hot water jug, followed by not undressing, etc. Initially they improvised a wood-burning stove made from a carbide can, but smoke from it engulfed the building with clouds of soot to the extent that soot dropped on their plates while they ate. Finally they acquired some stovepipe and vented the soot through the roof. This addition made it much more comfortable in the "summer house".

## Competition from Langley

The brothers' deepest concern was one over which they had no control. The Wrights were in a competition with Samuel Pierpont Langley, Secretary of the Smithsonian Institution, to be the first person in history to achieve manned, heavier-than-air, powered flight. It was a strange competition. Although the Wrights knew about Langley's work, he was completely unaware of their plans to build a powered, man-carrying machine.

Langley had obtained a $50,000 contract from the United States War Department to make a powered machine based on his successful, steam-engine-powered, un-manned airplane. The Wrights knew Langley had a marvelous 52 horsepower, lightweight engine that was much better than theirs. They also knew that Langley was due to test his plane in early October and that they could not hope to be ready that soon. So they decided, uncharacteristically, to fly their new Flyer as soon as was assembled. They would shun caution by canceling their usual plans to fly their large machine as a kite and as a glider.

Langley's plan was to launch his 48-ft craft by means of a catapult system mounted on a houseboat anchored in the middle of the Potomac River, FIGURE 77. He made his first attempt on October 7 with his assistant Charles Manly at the controls. With the engine roaring, the plane moved forward only to slide lazily into the water. Manley was not hurt. One reporter said the ungainly machine slid into the water like mortar slipping from a mason's trowel. Langley was mortified. He claimed that

Figure 77

the plane had caught on the catapulting mechanism and that the airplane design was sound. He vowed he would try again.

Wilbur was relieved when he heard of Langley's failure to fly. However, he did not know that Langley was planning to try again. In a letter to Chanute, he said:

*"I see that Langley has had his fling and failed. It seems to be our turn to throw now and I wonder what our luck will be."*

## Testing the New Machine

By mid October, the upper wing of the new machine had been assembled and covered. In a letter to his father, Orville wrote:

*"We completed the upper surface of our new machine yesterday. It is the prettiest we have ever made, and of a much better shape, being smooth on both upper and lower sides."*

He went on to say that since their old wings were better than their competitors, this wing represented a major advancement in aeronautical efficiency.

# Propeller Shaft Failure

On November 5th, the machine was essentially completed and ready for the first total power plant test. It took a number of attempts before the engine would run reasonably well. Still, the engine continued to misfire causing the propellers to vibrate so severely that almost immediately both propeller hubs broke loose where they were welded to the propeller shafts. This failure was a serious setback for the brothers. They had no alternative but to ship the shafts back to Charlie Taylor in Dayton for repair.

Just about this time, Octave Chanute arrived with the news that Langley was preparing to make a second trial of his machine in early December. More gloom descended upon the Wrights. It would be nearly impossible for them to beat that date.

Now the brothers were really discouraged. They were battling four demons simultaneously: Langley, the weather, mechanical failure, and now they were beginning to doubt their ability to succeed. The engine test had been so short that they had missed their first opportunity to take the measurements of the engine-propeller propulsive performance. These measurements were required to verify the accuracy of their theoretical propulsion calculations. Further, after assembling the big machine they found that it had grown in weight from their projected weight of 625 pounds to more than 700 pounds. Would they have enough power for flight?

With the project on hold it was a real challenge for the brothers to keep these demons at bay before the repaired propeller shafts arrived. It was to be a 15-day wait!

Octave Chanute's 6-day visit was unsettling for the brothers. Before he arrived, Wilbur and Orville were worried about the increase in the actual weight of the new machine, the failure of their propeller shafts and the certainty that Samuel P. Langley would fly before they did.

Chanute did not allay any of their fears. Instead, he told Wilbur and Orville that no one had designed a flying machine with such small margins of safety as theirs and that he disagreed with their estimate of a five percent power loss caused by friction in their chain drive system. Instead, he said that they should expect these losses to be at least 25 percent to 30 percent. He wondered if the propellers would receive enough power to attain flight.

The brothers were relieved when Chanute left camp. Unfortunately he left his doubts behind to smolder in their minds. At times they felt doomed to failure.

# Mounting Optimism

For relief from their worries, Wilbur and Orville busied themselves by conducting tests, making adjustments on the machine and going over their calculations.

They tested their "junction railroad" by launching their 1902 glider from the rails set on the incline of the Big Hill. The brothers would let the piloted glider roll down the inclined track. When flying speed had been attained they would use the front rudder control to lift the glider off the rail. They were pleased by their ability to make five out of six successful launches in this manner. However, the aging glider was not flying as well as usual. They found that their need to heat the hangar had dried out the wood and cloth to the point that the machine was no longer safe to fly.

Next they tested the ability of the new machine's front rudder to withstand strong wind loads, by taking it outdoors and holding it at various angles against a 33 mph wind. They found its structural strength was more than adequate.

They also tested the ability of the wing to withstand flight loads by suspending the entire airplane by its wing tips from the rafters of the hangar and applying a 440-pound weight. This weight was much larger than the wing was required to carry. Although the test was successful, they did notice that the fabric on the wing tips wrinkled badly under the load. By rearranging the control wires they eliminated this potential loss in aerodynamic efficiency.

Finally, the brothers conducted a power transmission efficiency test to allay their concern for Chanute's estimate of transmission power loss. A weight equivalent to that which the engine would exert on the chains was hung on a chain that was threaded over one of their sprockets. They were pleased to note that the force required to raise the weight indicated that indeed their transmission losses were only about five percent as they had predicted. Chanute was wrong! Things were looking up, but the most important question still remained. Did they have enough propulsive power for flight?

The repaired propeller shafts arrived about noon on November 20 and the propulsion system was ready to test that evening. Then another problem arose. The engine vibrated so severely that the sprockets on the propeller shaft continually became unscrewed. Try as they might they could not tighten the sprockets on the shaft sufficiently to prevent them from loosening during engine runs.

Sleeping on the problem, they decided to use bicycle technology for its solution. They spread Arnstein's hard tire cement on the screw threads, screwed on the sprockets, heated the assembly, and when it had cooled the sprockets were on so tightly that they wondered if they would ever be able to remove them.

Wilbur and Orville also discovered the source of their engine's rough running performance. Vibration had caused the fuel valve to slip resulting in an uneven flow of fuel. Repairing the fuel valve solved that problem. Now at last, on November 21st, they believed that they were ready to conduct their long-anticipated test of the entire propulsion system.

## The First Successful Propulsion System Test

The propulsion system test results were beyond the brother's expectations. Hoping for a propeller speed of at least 305 rpm, they were surprised when the engine delivered a propeller speed of 350 rpm during a 60 second run.

Afterward, they conducted a propeller thrust test. The machine was set on rollers. A rope was tied to the machine, strung over a pulley and tied to a box of sand. With the engine running, the propeller pushed the machine forward and the thrust force was determined by the weight lifted.

From this test the Wrights learned that their propellers were generating between 132 and 138 pounds of thrust at a propeller speed of 350 rpm. The brothers were elated! Their calculations had predicted a thrust of only 90 pounds and now they were getting 132 pounds. All of their doubts had been overcome. This extra amount of thrust would more than offset the machines weight increase.

Finally, they tested the strength of the wing with the engine running. Again, the machine was suspended by the wing tips inside the hangar. This time a pilot was aboard and the engine was running. No problems occurred and the test proved that the in-flight strength of the machine was satisfactory. With only a few minor adjustments, they would be ready to fly.

Then on November 28th, disaster struck again. After conducting a few short engine runs, they suspected that something was wrong. Upon inspection, they found hairline cracks in one of the propeller shafts.

## Second Propeller Shaft Failure

Again, despair set in. Langley was due to fly at any moment and they had no alternative but to send Orville back to Dayton to obtain stronger propeller shafts. This time new shafts would be made from solid spring steel instead of hollow steel tubing.

Figure 78 Courtesy of Carillon Historical Park Archives

Orville left for Dayton on November 30 and returned to camp with the new shafts on the evening of December 12. He also brought news that Langley had attempted to fly on December 7 and this time he had failed catastrophically, as shown in FIGURE 78. With his financial resources exhausted, Langley could not try again.

Relieved that Langley was out of the "race," the brothers installed the new propeller shafts and were ready for a flight test that night. However, the wind wasn't strong enough for flight. Instead, they tested the launching system by running the machine along the track under its own power with a pilot aboard and the other brother running alongside to steady the craft. On one of the runs, the bottom, rear, rudder support caught on the track and was broken. Fortunately, only minor repairs would be required.

The next day Sunday, December 13 was a warm, beautiful day with the wind blowing about 18 miles per hour. This was a perfect day for the first flight. There was even enough wind for a launch from level ground.

But no, the brothers did not fly! It was the Sabbath. Wilbur and Orville had promised their father that they would observe the Sabbath; they always had, even here at Kitty Hawk. Can you imagine the temptation they must have felt? Here at long last they had the machine of their dreams, it was working perfectly and the weather was perfect. Still they were true to their word and beliefs. Never on Sunday for the Wrights!

### First Flight Attempt

December 14 was another beautiful day. The Wrights spent the morning repairing the tail support structure damaged two days earlier. At one o'clock they put out a signal flag to notify the men at the Kill Devil Life Saving Station that they were ready to fly. These men had volunteered to help the Wrights during their flight trials. Since the wind speed was only 5 mph, the Wrights knew they would not be able to take off from level ground as they had planned. Instead, they would have to launch on a down hill track in order to get a boost from gravity.

Five men and two young boys answered their call for help to drag the machine to the Big Hill for the trial.

They laid the rail on a 9-degree slope of the hill and put the machine on the track, as shown in FIGURE 79. The minute the engine sprang to life the two boys "headed for the hills" in fright. They were unaccustomed to such a noisy engine. The

FIGURE 78
Langley's second crash and last attempt to fly, December 8, 1903.

FIGURE 79
1903 flyer on slope of big hill ready for Wilbur's flight attempt on December 14, 1903.

Figure 79 Courtesy of Special Collections and Archives, Wright State University

brothers flipped a coin and Wilbur won the chance to become the first man in history to fly a powered machine.

With the engine roaring, Wilbur released the restraining cable and the craft started down the track with Orville steadying it at the right wing. After 35 feet, Orville could no longer keep up. He released his hold and Wilbur pulled back on the front rudder control to climb into the air.

And climb he did! Wilbur found that he was flying something significantly different from a glider! The new machine's extra power and speed made its control response much more sensitive than that of the glider. It climbed rapidly nose high until it was 15 feet in the air. Then the machine began to lose speed and Wilbur tried to bring the nose down. However, the machine had stalled and slowly lost altitude with the left wing low. The wing tip struck the ground first and the machine spun around breaking one of the skids as well as parts of the front rudder structure. It was a soft landing and Wilbur was unhurt. He had traveled about 112 feet from the end of the track, FIGURE 88. The Wrights certainly did not consider this to be an official, sustained flight.

Aside from his disappointment in his error in maneuvering the machine, Wilbur was pleased. The landing was nice and easy for him, everything worked well, and there was enough power to fly. All they needed to do was learn how to fly it. Repair of the broken parts would be simple.

Wilbur sent a telegram to his father saying:

> *"Misjudgment at start reduced flight one hundred twelve power and control ample rudder only injured success assured keep quiet."*

While repairs were being made, the brothers watched two beautiful days pass. They were ready to fly again on December 16, so they took the machine out for a second try. However, the wind was not strong enough, so the flight attempt was postponed until the next day.

They went to bed hoping for stronger winds when they awoke. They were not to be disappointed.

FIGURE 80
Wilbur's hard landing 112 feet from the end of the track.

# Chapter 11 – "Success, Four Flights"

A major change in the weather pattern came to Kitty Hawk shortly after midnight on the morning of December 17. The balmy weather of the past few days was driven out by a fierce 'nor'easter' that brought winter back with all of its fury. Temperatures plummeted and winds howled across the land driving the wind chill down to nearly 0 degrees Fahrenheit. By the time Wilbur and Orville arose, ice had formed in rain puddles and they were measuring wind speeds of 24 to 27 mph. The weather was so bad that very few birds were flying at a time when they usually filled the air. The Wrights did not take the hint!

With no abatement of the wind by 10 o'clock, the brothers decided to attempt a flight in spite of the dangerous weather conditions. They hung out the flag to signal the men at the Life Saving Station, trundled their machine out of the hangar and laid out the track on a smooth, level stretch of ground about 100 feet north of the hangar. The biting wind made work difficult and the brothers had to retreat to their "summer house" several times to take advantage of the warm fire that was blazing in their stove.

Five people came to their aid: John Daniels, W.S. Dough and A.D. Etheridge from the Life Saving Station; W.C. Brinkley of Manteo and a teen-aged boy, Johnny Moore, of Nags Head. An invitation had been given to the general community. However, few were willing to brave the elements just to see "another flying machine not fly." Since Wilbur had attempted to fly three days earlier, it now was Orville's turn.

Orville set up a camera and called John Daniels to take a picture of the flight, FIGURE 81. He might have said something like this:

*"Mr. Daniels, I have focused this camera on the front of the track. I will get on the machine, turn on the engine and release the machine to start the flight. When the machine arrives at the end of the track, I want you to squeeze this bulb to take the picture no matter what happens." And John Daniels most assuredly replied, "Yes Mr. Wright, I will do that."*

Then the two brothers walked to the side of the craft and spoke quietly to each other. Daniels later described that conversation:

*"After a while they shook hands - and we couldn't help note how they held on to each other's hand, sort of like they hated to let go; like two folks parting who weren't sure they'd ever see each other again. Wilbur took position at the right wing tip and then motioned to the five onlookers, urging them not to look sad, but to laugh, [shout] and clap our hands and try to cheer Orville up when he started."*

## KEY NUMBER TWENTY-ONE •
## THE WORLD'S FIRST MANNED, HEAVIER THAN AIR, POWERED CONTROLLED, SUSTAINED FLIGHTS

At 10:35 a.m. with the engine running at full power, Orville released the restraining wire and the machine rolled forward against the gale with Wilbur steadying it on the right. After a 40-foot run Orville raised the front rudder and the machine quickly left the ground. Immediately Orville realized he had a wild mustang by the tail. The machine darted up and down like a roller coaster. After traveling about 100 feet, the machine jumped to an altitude of 10 feet and upon Orville's correction, it suddenly darted downward and struck the ground about 120 feet from the end of the starting

FIGURE 81
Illustration of Orville instructing John Daniels about taking a picture of the first flight on December 17, 1903.

Figure 81

rail. Twelve seconds had elapsed since he left the ground. The length of the flight was equivalent to a 540-foot flight in still air.

Amid cheers and congratulatory handshakes, the machine was dragged back to camp. The Wrights invited everyone into their living quarters to get warm while they repaired the front skid, which had been slightly cracked.

And this is the picture that John Daniels took! This first flight picture, FIGURE 81, is the most significant and one of the most thrilling aeronautical pictures ever taken. There, artfully composed, we see the 1903 Wright Flyer in full flight before it has reached the end of the track. Wilbur has released his hold and Orville has inclined the front rudder significantly upward to lift the machine into the air. There we can see Wilbur's footprints as he ran alongside the machine during the take off run as well as the bench and C-clamp used to support the wing prior to flight. And finally, we can see the spade used for leveling the ground while laying the track, the can of grease for lubrication of engine and transmission parts

and the box carrying the battery and spark-coil used to start the engine.

Wilbur was the pilot for the second flight, which began at 11:20 a.m. His flight also was a roller coaster ride but he flew out a little farther, about 175 feet, also in about 12 seconds.

The third flight began at 11:40 a.m. with Orville piloting. He progressed the furthest yet until a gust of wind struck him from the right side and raised the right wing abruptly. Orville quickly responded with the wing warping control, but he over-controlled to the point that the right wing rolled downward and struck the ground as shown in FIGURE 83. Orville had traveled about 200 feet in 15 seconds.

At noon, Wilbur took the controls for the fourth flight. His flight began in the usual oscillatory fashion for the first 200 feet. Then, the oscillations became even more vicious as the machine dived downward nearly out of control. But Wilbur pulled it out a scant one foot above the ground and continued the flight. As the machine approached the 300-foot mark, the oscillations began to smooth out with

FIGURE 82
The first sustained, man-carrying, heavier-than-air, powered, controlled flight in the history of the world. On December 17, 1903 Orville Wright flew 120 feet in 12 seconds.

Figure 83 Courtesy of Special Collections and Archives, Wright State University

FIGURE 83
Third flight by Orville Wright, about 200 feet in 15 seconds.

FIGURE 84
Fourth and longest flight by Wilbur Wright: 852 feet in 59 seconds.

the machine well under control while undulating between eight and fifteen feet above ground. Wilbur was really flying! He was careful not to exceed the 15-foot, safe-altitude limit that he and Orville had agreed upon, which made flying more difficult. But as usual, the practice of safety prevailed. As the flight distance increased to 400, 500, 600 and 700 feet, it appeared that Wilbur was well on his way to Kitty Hawk, four miles away. And indeed that was Wilbur's objective. As he passed 800 feet a little hummock appeared before him and he eased the control to increase his altitude. But swirling turbulent eddies around the hummock controlled the flight's destiny as they struck the craft and caused it to begin pitching again. Control again became difficult and finally the machine dived violently emitting a splintering noise as it struck the ground. Wilbur had flown a distance of 852 feet for flight duration of 59 seconds. This flight was equivalent to flying one-half mile in still air. FIGURE 84, taken from the end of the launching rail, shows the Flyer at the end of the fourth flight.

The four flights made that day were collectively the twenty-first Key to Wright Brothers' Success.

Figure 84 Courtesy of Special Collections and Archives, Wright State University

FIGURE 85
Damaged front rudder after Wilbur's long flight.

As the men rushed to congratulate Wilbur, they noticed the front rudder was drooping because the rudder braces had been broken during the crash, FIGURE 85. This minor damage could be repaired easily. They removed the front rudder and dragged the machine back to camp on its landing skids, a pleasant task with the strong wind at their backs and their exhilaration from the tremendous flight that had just been made.

## Upset!

As they were standing around the machine planning for repairs and additional flights, Mother Nature intervened, as shown in FIGURE 86. A strong gust of wind hit the Flyer from the side and the men scrambled to hold it down. But the gust was too strong. Daniels was thrown into the space between the two wings and the Flyer cart wheeled wing-tip-over-wing-tip like tumbleweed in a Kansas windstorm. The others were thrown about and Daniels was tangled amid the chain guides, engine, wires and breaking parts. Daniels referred to this as "my first - and God help me - my last flight."

He continues:

*"I found myself caught in them wires and the machine blowing across the beach and heading for the ocean, landing first on one end and then on the other, rolling over and over, and me getting more tangled up in it all the time. I tell you, I was plumb scared. When the thing did stop for half a second I nearly broke up every wire and upright getting out of it."*

Miraculously, his injuries were only temporary dizziness and minor cuts, scrapes, and bruises!

Upon inspection it was obvious that no more flights were going to be made that year. In fact, the 1903 machine never flew again. The cast aluminum engine mount was broken, chain guides were severely bent, and many major structural components were severely damaged. There was nothing to do but break camp and return to Dayton.

After noon-time dinner, the brothers walked to The Kitty Hawk Weather Station to send the telegram shown in FIGURE 87. Mistakes were made in the transmission: 57 seconds for the longest flight instead of

59 seconds, and they misspelled Orville's name.

Of course the Wright family was excited with the news of the flights. But Wilbur and Orville's nieces and nephews had a different view. The children's response to the telegram according to their niece, Ivonnette Wright Miller, was something like: "Oh, goody, Uncle Will will be home in time to carve the Christmas turkey!" He performed this traditional duty with a gusto that was very entertaining for the young Wrights.

Looking back on that stormy day in December 1903, Orville later wrote:

*"With all the knowledge and skill acquired in thousands of flights in the last ten years, I would hardly think today of making my first flight on a strange machine in a twenty-seven mile wind, even if I knew that the machine had already been flown and was safe. After these years of experience, I look with amazement upon our audacity in attempting flights with a new and untried machine under such circumstances!"*

In December 1913, Orville described his first flight and in doing so provided the definition of powered airplane flight for comparison with the claims of other contemporary would-be aviators.

*"This flight lasted only 12 seconds, but it was nevertheless the first in the history of the world in which a machine carrying a man had raised itself by its own power into the air in full flight, had sailed forward without reduction of speed and had finally landed at a point as high as that from which it started."*

In view of Wilbur's fourth flight, Orville could have added that the Wright's flights also were made under control. No flight by anyone who claimed to have flown before the Wrights could live up to this definition.

The Wrights knew what they had accomplished. In a January 1904 news release to the Associated Press, Wilbur wrote that once they realized they had achieved all of their 1903 flight objectives at Kitty Hawk,

*"We at once packed up our goods and returned home, knowing that the age of the flying machine had come at last."*

Figure 86

FIGURE 86
Flyer upset and severely damaged by a strong gust of wind.

FIGURE 87
Telegram to Bishop Wright after four successful flights at Kitty Hawk

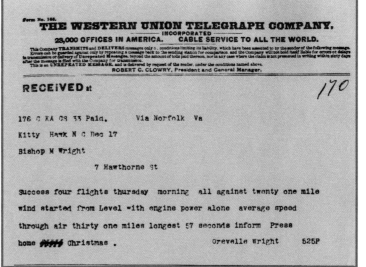

# Chapter 12 – A Practical Flying Machine

Wilbur and Orville Wright were at a crossroad in their lives. They could either continue aeronautical experimentation as an avocation or become aeronautical businessmen. They decided to travel the business route.

They didn't waste any time. In early January 1904, the brothers began construction of a new engine and a new flying machine, the 1904 Flyer. They also retained the services of H. A. Toulmin, a patent attorney in Springfield, Ohio, to file both foreign and domestic patents on their invention.

The Wrights had decided it would be more efficient to conduct all further flight experiments in Dayton. Soon they had obtained permission from Torrence Huffman, a west-side banker, for the rent-free use of his 100-acre pasture located next to Simms Station on the D.S. & U. interurban traction line eight miles northeast of Dayton, FIGURE 88. Huffman permitted them to build a hangar there and his only stipulation was that they must drive the cattle and horses outside the fence every time they flew. By the end of April they had erected a hangar and the new airplane was taking shape.

## 1904 Flyer Flight Trials

The new 1904 Flyer, FIGURE 89, looked much like the 1903 machine. However, it was stronger, weighed nearly 900 pounds with pilot and had 77 pounds of ballast added in front in order to improve the pitch (nose up and down) stability. The machine was equipped with a new, more powerful engine that provided from 15 to 16 horsepower.

The Wrights invited the area newspapermen to witness the first 1904 flight attempts on the May 23 and 26. No picture taking was permitted. Unfortunately, the machine did not fly on the first day and flew only about 25 feet on May 26 because of light winds and poor

engine performance. The reporters became disinterested and did not return.

The brothers' difficulty in taking off plagued them most of the summer. By mid August, they had not even exceeded the 852-foot Kitty Hawk flight. Their "junction railroad track" lengths grew to 236 feet as they struggled to rise from the ground in the weak, shifty winds prevalent in the Dayton area at that time of year.

The Wrights' takeoff problem was not unique. For example, modern aircraft face the same difficulty when taking off from Denver. Colorado, "the mile high city," on a hot summer day. The 'hot Denver day' condition causes a lower-than-normal air density and hence produces lower-than-normal lift and propulsive force. These low air density conditions cause a much longer ground run for the plane to acquire the higher than normal ground speed needed for takeoff. If the runway is not long enough, the pilot must board fewer passengers, carry less freight, wait for a stronger headwind or delay the flight.

The Wrights experienced two different flying conditions in 1903 and 1904: i.e. 34 degree Fahrenheit air at Kitty Hawk at sea level versus 81 degree Fahrenheit air at Huffman Prairie, which is 815 feet above sea level. The difference between the two conditions resulted in the equivalent of a 4700-foot temperature/altitude reduction in air density, as shown in FIGURE 90. The end result was a 13 percent reduction in lift as well as a comparable reduction in propeller thrust for the Huffman Prairie trials compared to that experienced at Kitty Hawk in December 1903.

Compounding the density problem, the wind in Dayton on May 23,1904, was only 3 mph compared to the 27 mile per hour gale at Kitty Hawk. If the Wrights needed to achieve a flying speed of 30 mph for takeoff, they would have had to accelerate to a track speed of only 3 mph at Kitty Hawk compared to a high track speed of 27 mph at Huffman Prairie because of the difference in wind speeds. Computations made by Harry Combs for his book, *Kill Devil Hill*, indicated that 1904 Flyer, with its small engine, would have needed a 750-foot track to reach a 30 mph takeoff speed on

FIGURE 88
Map of Huffman Prairie and surrounding area.

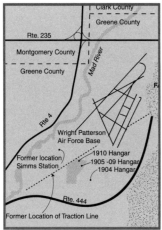

Figure 88

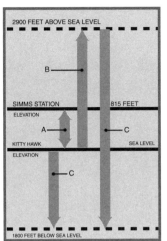

Figure 90

Figure 89 Courtesy of Special Collections and Archives, Wright State University

that day in May, and yet their track that day was only 100 feet long.

A catapult, shown in FIGURE 91, solved the problem. A tall tower that the Wrights called a "derrick" was placed behind the track and a large weight, eventually increased to 1600 pounds, was hoisted to the top. A rope tied to the weight was strung around pulleys at the top and bottom of the tower and led to a pulley at the front of the track. From there the rope was led back to and looped over a hook on the airplane, which was held in position by a restraining wire. With engines running full speed, the pilot would release the restraining wire and the airplane would receive a powerful forward pull as the weight fell. The rope automatically slid off the hook on the airplane as it rose from the track.

This "derrick" device worked perfectly and became the standard launching procedure for the Wrights for many years. Use of the catapult enabled the brothers to reduce the "junction railroad" track length to about 80 feet.

Thereafter, flight times soared. On September 20, Wilbur completed the first circle of the field with a flight of 4080 feet in 1 minute, 35 seconds. Before the year ended both men were circling the field and both had logged flights of nearly 3 miles with durations of about 5 minutes. The 1904 Flyer making a flight on November 16 is shown in FIGURE 92.

Although 105 flights had been logged in 1904, the total duration of these flights was

only about 45 minutes. Because of its many crashes, the 1904 Flyer had been thoroughly worn out and was not saved.

## 1905 Flyer Flight Trials

Wilbur and Orville built a completely new flying machine for the 1905 flight trials. A three-view drawing of the 1905 Flyer or the Flyer III, as many know it, is shown in FIGURE 93. Changes from the previous year's model included enlargement of the front and rear rudders and extending the rear rudder further behind the wing to improve controllability and pitching stability. Blinkers, semicircular vertical surfaces, were installed between the front rudder surfaces to improve stability in a turn. The entire structure also was strengthened. A major change was made in the control system. Recall that since 1902 the brothers had flown with the rear rudder directly connected to the wing warping mechanism. In 1905 they disconnected

FIGURE 89
Orville and Wilbur discussing plans for their new 1904 flyer at Huffman Prairie.

FIGURE 90
Air density problem at Huffman Prairie during the summer of 1904.

FIGURE 91
Catapult system devised to overcome launching problems at Huffman Prairie.

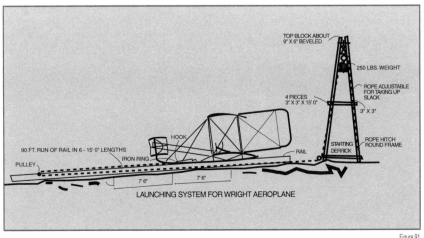

Figure 91

Figure 92 Courtesy of Special Collections and Archives, Wright State University

FIGURE 92
  1904 Flyer in flight,
November 1904

FIGURE 93
  Three-view drawing of
1905 Flyer

FIGURE 94
  Bent end propeller used
on 1905 Flyer.

these two controls and provided one separate activation device for each. A lever at the right hand of the pilot activated the rear rudder control surface and the hip cradle was retained to activate the wing warping system. As before the pilot's left hand operated the front rudder control lever. This separation of the control devices gave the pilot complete freedom to operate

the control system in any type of flight situation

The 1904 engine was reused on the 1905 machine. By this time the 1904 engine's power had increased to about 20 horsepower simply because it had become well-broken-in by that time. These changes increased the overall 1905 Flyer weight to about 925 pounds, including the pilot and about 70 pounds of ballast for improved pitch stability.

The Wrights had noticed in 1904 that the thin tips of their propellers were twisting under load causing a loss in efficiency. To combat this problem in 1905, they designed what became known as their bent end propellers. These propellers were formed by cutting a pie-shaped segment out of the outer section of the blade's leading edge as shown in FIGURE 94. This design concept solved the problem and was the standard type of propeller used for Wright airplanes for years to come.

Work on the 1905 machine began at Huffman Prairie about May 23 and the first flight, made with the catapult, occurred on June 23 as shown in FIGURE 95.

Unfortunately, the brothers' initial 1905 test results were equally discouraging as those in 1904. During the first 21-day period only eight flights were made and the longest, 774 feet, did not equal the 1903 Flyer's fourth flight at Kitty Hawk. Considerable difficulty was caused by control problems, pilot error and engine performance. There were frequent crashes followed by time-consuming repairs.

Orville's crash on July 14 was the type accident that the brothers had always feared. After seeming to fly correctly, the machine suddenly began undulating and then dived and hit the ground at 30 mph, bouncing three times. The crash was devastating. Orville was sent flying out through the upper, broken surface of the wing. He came to rest dazed, bruised but uninjured. Damage to the Flyer was extensive.

At that point the Wrights vowed to improve the pitch stability while they rebuilt the machine. They increased the size of the front rudder and placed it more than 4 feet further in front of the wing. These

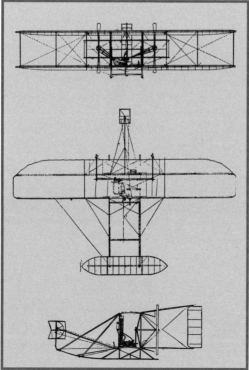

Figure 94        Figure 93 Courtesy of Special Collections and Archives, Wright State University

modifications increased the weight by about 15 pounds and made the 1905 Flyer seven feet longer than the 1903 Flyer.

## KEY NUMBER TWENTY-TWO •
## LEARNED HOW TO FLY

The first flight of the modified 1905 Flyer was on August 24. From that time on the brothers' progress was phenomenal. Flights of 1556 feet and 2263 feet were made that first day, and by the end of August, flight times had increased beyond one minute. By mid- September, flights of 4 and 5 minutes were becoming common. On September 8, Orville flew a full circle of the field followed by a figure eight during the same flight! The new design combined with their improved piloting skill was paying off. But the best was yet to come.

On September 26, Wilbur flew for 11 miles in 18 minutes and for the first time in 1905 had to land because the plane had run out of gas. Flight durations continued to soar. On October 4, Orville flew for 20.8 miles. Then, on October 5, Wilbur made nearly 30 circuits of the field and established their record flight: 24.2 miles in 38 minutes on a flight like the one shown in FIGURE 96. Again he had to land because of an empty gasoline tank. Having achieved complete control of their flying machine, the Wrights had finally learned how to fly. This accomplishment became the twenty-second Key to the Wright Brothers' success.

## KEY NUMBER TWENTY-THREE •
## DEMONSTRATED A PRACTICAL
## AIRCRAFT

The October 5 flight was particularly remarkable because its duration was nearly equal to the total flight time logged by the brothers during 1903 and 1904. Although only 49 flights were flown in 1905, their total flight time in the air had jumped to about 3.7 hours.

With an airplane that was able to travel significant distances and to land without being wrecked, they now had a practical product to sell to the world! They were ready to go into business. Development of the first practical airplane was the twenty-third Key to the Wright Brothers success.

Figure 95 Courtesy of Special Collections and Archives, Wright State University

FIGURE 95
First flight of 1905 Flyer, June 23, 1905. Note the catapult launching device and hangar building.

FIGURE 96
1905 Flyer flying on October 5, 1905 at Huffman Prarie.

Figure 96 Courtesy of Special Collections and Archives, Wright State University

# Chapter 13 – Final Triumph

The completion of the 1905 flight trial marked the end of Wilbur and Orville's brilliant scientific developments. Of course, they would make improvements to their Flyers, but from here on they would concentrate their efforts on making a success of an airplane business.

## *Patents and Licensing*

Being patriotic citizens, the Wrights initially offered to build one airplane for the Army, but the Army was so disinterested that the offer was rejected out of hand, apparently without reading the offer. Since they wanted to sell licenses and airplanes, the Wrights reluctantly went to Europe in search of customers.

The Wrights' United States patent, shown in FIGURE 97, was granted on May 23, 1906. The patent was based on their control system, and the claims were so broadly written that anyone who wanted to build a controllable airplane would have to come to the Wrights for a license. Foreign patents also were granted in Austria, Belgium, France, Germany, Great Britain and Italy. But the brothers soon were to learn that it is one thing to obtain a patent--- another thing to sell it or license it.

For two years the brothers promoted their airplane in the United States and in Europe. Some apparently good prospects did not materialize and other potential customers were really trying to steal their secrets. The brothers seemingly made it hard on themselves by not allowing anyone to see their airplane without having signed a firm contract. However, they believed their airplane was so simple that anyone with a camera and measuring tape could copy it. Disillusioned by this foreign deception, the honorable Wrights returned to the United States wondering if they would ever succeed in this business. And then their fortunes turned.

In early 1908, they received contracts from the US Army Signal Corps and from a syndicate in France, known as La Compagnie Generale de Navigation Aerienne (CGNA). The Army contract price was $25,000 for one aircraft with a speed bonus and the CGNA contract promised 500,000 francs for the first machine, a 50% share of the founder's share of stock, and 20,000 francs for each of four additional airplanes delivered. Both contracts required flying demonstrations in the fall of 1908 before final purchase.

## *Kitty Hawk, 1908*

The Wrights decided to return to Kitty Hawk for flying practice in order to take advantage of the strong steady winds there. They took their trusty 1905 Flyer, which had been modified to accommodate a pilot and a passenger seated in an upright position. With no hip cradle, a completely new system of levers was needed to

FIGURE 97
United States Patent on Wright control system issued May 23, 1906

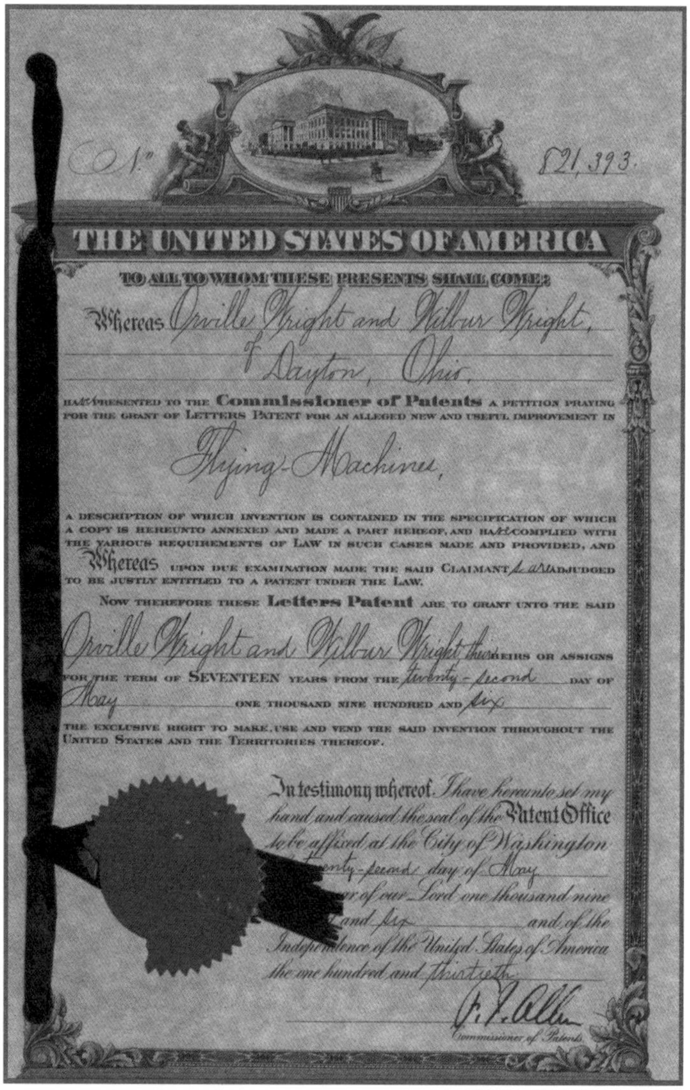
No 821,393.

THE UNITED STATES OF AMERICA

TO ALL TO WHOM THESE PRESENTS SHALL COME:

Whereas Orville Wright and Wilbur Wright, of Dayton, Ohio,

have presented to the Commissioner of Patents a petition praying for the grant of Letters Patent for an alleged new and useful improvement in

Flying-Machines,

a description of which invention is contained in the specification of which a copy is hereunto annexed and made a part hereof, and have complied with the various requirements of Law in such cases made and provided, and

Whereas upon due examination made the said Claimants are adjudged to be justly entitled to a patent under the Law.

Now therefore these Letters Patent are to grant unto the said Orville Wright and Wilbur Wright their heirs or assigns for the term of Seventeen years from the twenty-second day of May one thousand nine hundred and six the exclusive right to make, use and vend the said invention throughout the United States and the Territories thereof.

In testimony whereof I have hereunto set my hand and caused the seal of the Patent Office to be affixed at the City of Washington ... twenty-second day of May ... of our Lord one thousand nine ... and six and of the Independence of the United States of America the one hundred and thirtieth.

Commissioner of Patents

Figure 97 Courtesy of Carillon Historical Park Archives

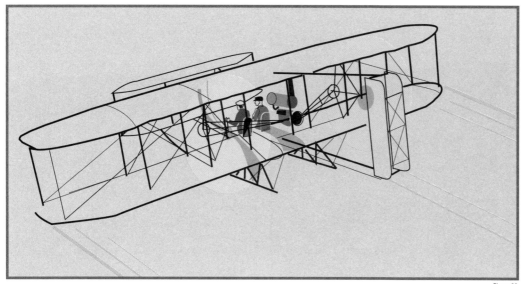

Figure 98
  Illustration of Charles
Furnas flying as passenger
with Orville at Kitty Hawk
May 14, 1908

Figure 98

actuate the Wrights' usual control surface operations. The modified aircraft also had a new engine that produced about 32 horsepower, which was about fifty percent more powerful than the original 1905 Flyer engine. Charlie Furnas, a friend of theirs who was a mechanic, arrived unannounced at the Kitty Hawk campsite and offered to help with the flying preparations. He was rewarded on May 14, 1908, when he became the first person to ride as a passenger in a Wright airplane as seen in Figure 98. After logging only 15 minutes of short flights over a 9-day period, Wilbur lost control and crashed at high speed. The plane was demolished and repairs were beyond the question. Miraculously, Wilbur was not injured.

Since the Wrights were required to perform on both contracts simultaneously, and since neither party would accept a delay, the Wrights reluctantly divided forces for the first time. Wilbur went to France and Orville went to Ft. Meyer, Virginia, to fulfill the two contracts.

## Wilbur at Le Mans, France

Wilbur traveled to Le Mans, France, where he was provided an assembly space in a small building that was part of an automobile factory owned by Leon Bollèe, one of his promoters. Wilbur literally lived and worked in that small building.

Wilbur was shocked when he opened the crates containing the airplane parts. French

customs officials, while unpacking and repacking the crates during the inspections had damaged many of the parts. An engine previously sent over also was damaged. Finding no workmen with dependable skills who also could speak English, Wilbur faced a major rebuilding job alone. He realized he had no room for error. He had to prove conclusively that the airplane could fly the first time he tried in order to succeed in his venture. Consequently, Wilbur proceeded very carefully and slowly to make sure everything was done correctly.

In his careful methodical manner, Wilbur took his time assembling the plane in spite of ridicule from newspapers. He was called a bluffer and some people cried, "The Wrights are not fliers, they are liars!" At that time, French aviators had been gradually learning how to make airplanes fly in straight lines and in some cases skid around corners. They thought they were the best fliers in the world.

Wilbur's nearly assembled plane was taken to the racetrack at nearby Les Hunaudières where Wilbur made final preparations for the flight.

All was ready on August 8, 1908. With the Flyer attached to the catapult, Wilbur was ready, Figure 99. French aviators, Louis Blèriot, Paul Zens and Ernest Archdeacon were among the small gathering of people expecting to witness Wilbur's certain failure.

But Wilbur really showed them who

Figure 99 Courtesy of Special Collections and Archives, Wright State University

Figure 100 Courtesy of Special Collections and Archives, Wright State University

could fly that day! In the late evening, Wilbur released the restraining cable and he was off on a flight that the French would never forget, FIGURE 100. He flew straight ahead, and when it appeared he would hit the trees at the end of the track, he banked sharply, and the people gasped expecting to see an accident. But no, Wilbur pulled the Flyer out smoothly and continued smoothly around the track under perfect control. The people had never seen a plane bank like that. Wilbur made two complete circles around the field and ended the flight coasting in for a smooth landing. Everyone ran excitedly to greet him. "They are fliers, not liars," became the cry. The French aviators said, "We are but babes compared to the Wrights. We are beaten!"

Wilbur became an instant celebrity. Impressed by his modesty, sincerity and honesty, the flight was the beginning of a sincere love affair between the French people and the Wright family.

## Orville at Ft. Meyer, Virginia, 1908

Meanwhile, Orville was fulfilling his contract by showing the Army that he could disassemble and transport his airplane on a wagon, FIGURE 101. Then he began a series of flight tests and familiarization flights with a new airplane that he had never flown. Everyone was excited to see at last that indeed the Wrights could fly. The newspapers were having a heyday reporting first on a flight of Orville's and then on a flight made by Wilbur in France. As time passed, both were flying longer and longer. Finally Orville made a world record flight of 1 hour, 5 minutes. He was ready to perform the duration task required by the contract: carry a passenger aloft continuously for one hour.

Orville scheduled the duration flight on September 17, 1908. In FIGURE 102 he can be seen with Lt. Thomas E. Selfridge, the Army's designated passenger/observer. Orville took off and flew a few circuits of the field. Suddenly, he heard a ticking noise followed by a loud 'bang'! A propeller had split, caused a vibration and had cut a guy wire that held the rear rudder in proper alignment. The rudder twisted and forced

Figure 101 Courtesy of Special Collections and Archives, Wright State University

Figure 102 Courtesy of Special Collections and Archives, Wright State University

Figure 103 Courtesy of Special Collections and Archives, Wright State University

Figure 99
Wilbur ready to fly at Le Mans, France

Figure 100
Wilbur flying in front of grandstand of Les Hunaudières Racetrack, Le Mans, France, August 8, 1908

Figure 101
Orville transporting disassembled flyer on Wagon at Ft. Meyer

Figure 102
Orville and Lt. Thomas E. Selfridge, Army observer, ready for duration flight on September 17, 1908

Figure 103
Fatal crash during duration flight in which Lt. Selfridge was killed. He was the first person killed in a powered airplane.

Figure 104 Courtesy of Special Collections and Archives, Wright State University

Figure 105 Courtesy of Special Collections and Archives, Wright State University

After learning that Orville was going to heal, and refraining from flying a few days out of respect for Selfridge, Wilbur took to the air to demonstrate the airplane is still a wonderful invention even if accidents do happen.

Wilbur broke one record after another. FIGURE 104 shows him making a 115-meter altitude record by flying over a tethered balloon. Toward the end of December 1908, Wilbur had established a world flight duration record of 2 hours, 20 minutes.

FIGURE 105 demonstrates the enormity of the Wrights' advancement in the field of transportation, a Wright airplane flying over ox carts.

Wilbur moved his base of operations, first to a larger field located 11 kilometers east of Le Mans, Camp d'Auvours. Then as winter approached he moved to Pau, a luxurious resort area in southern France, where the weather was warmer. After Orville had recovered sufficiently from his injuries, he and his sister, Katharine, joined Wilbur in Pau, where the three can be seen in FIGURE 106 walking down the street dressed in their finery. The French people as well as royalty and the wealthy became as fond of the other two Wrights as they were of Wilbur.

Dignified and wealthy guests flocked to see him fly. King Alfonso VIII of Spain, shown with Wilbur in FIGURE 107, came to learn about the plane. But the queen wouldn't let him take a ride. Edward VII, King of Great Britain, also came to see Wilbur fly, FIGURE 109. Wilbur wrote:

*"There are more kings, queens, and*

FIGURE 104
Wilbur setting a world record 115-meter altitude record in France

FIGURE 105
Enormous advancement in the field of transportation: Wilbur flying over ox carts in France

the plane into a sharp dive. Orville tried to correct the flight path and almost made it. However the plane hit the ground hard with a loud crash, FIGURE 103. Both men were thrown from the wreckage. Selfridge died that evening, without regaining consciousness. Orville suffered fractures of his left thigh, several of his ribs, and injured his back.

## Back in France

Wilbur was horrified and grieved over Selfridges' death. Selfridge had entrusted his life to the Wrights and they had failed him.

Figure 106 Courtesy of Special Collections and Archives, Wright State University

Figure 109 Courtesy of Special Collections and Archives, Wright State University

FIGURE 106
Wilbur, Orville and Katharine, dressed in their best, reunited in Pau, France

FIGURE 107
King of Spain, Alfonso VII, shows his interest in the Wright Flyer

FIGURE 108
At Centocelle, Italy, King Victor Emanuel visits the flying field to watch Wilbur fly.

FIGURE 109
Edward VII, King of Great Britain, comes to Pau to meet the Wrights and see Wilbur fly.

Figure 107 Courtesy of Special Collections and Archives, Wright State University

Figure 108 Courtesy of Special Collections and Archives, Wright State University

Figure 110 Courtesy of Special Collections and Archives, Wright State University

FIGURE 110
Dayton's Wright Brothers Home Day Parade, June 17, 1909, honoring the world renown Wrights on their return from Europe .

FIGURE 111
2500 children forming an American flag at the Wright Brothers Home Day Celebration.

*millionaires here than a dog has fleas."*

From Pau they went to Centocelle, southeast of Rome, Italy for additional demonstrations of their flying machine. While there, King Victor Emanuel, FIGURE 108, witnessed some of their flights. It was a glorious time, that winter and spring of 1909. The Wrights returned by way of Great Britain to make arrangements for licensing and sale of airplanes there.

## Homecoming Celebration

The three Wrights' entrance into New York harbor on their return to the United States was overwhelming. As their ocean liner, the Kronprinzessin Cecile approached the harbor; it was met and escorted by an armada of boats of every size and kind. As they proceeded, flags were dipped, onlookers cheered, and whistles were blown as salutes to the victorious Wrights returning to their homeland. After subsequent festivities in

New York, Katharine, Orville and Wilbur returned to Dayton where they were greeted by the accolades of thousands of celebrating citizens. .

But that celebration was only the beginning. Shortly after their arrival, the city fathers visited the Wright home and informed the brothers of plans that had been made for a Wright Brothers Home Day celebration on the 17th and 18th of June. There were to be fireworks, parades, medal presentations and speeches for the two-day celebration. Wilbur and Orville were not impressed. Orville wrote:

> *"The Dayton presentation has been made the excuse for an elaborate carnival and advertisement of the city under the guise of being an honor to us. As it was done in spite of our known wishes, we are not as appreciative as we might be."*

FIGURES 110 and 111 show the parade and the celebration at the fairgrounds, where 2500 school children formed an American Flag while seated in the grandstand. But the Wrights had more important things on their minds. They were building a new version of their Flyer to use in a second try to fulfill the Army contract. The Army had given them a one-year extension to fulfill contractual obligations.

## Ft. Meyer, 1909

The 1909 Army Signal Corps Wright Flyer was specifically designed to exceed the Army's requirement for an airplane having a maximum speed of 40 mph. The basic contract price was $25,000. Also, there was a bonus of $2500 for each one mile per hour that the Wrights exceeded the 40 mph level. There was a compensating deduction of $2500 for everyone mile per hour less than 40 mph. If the plane did not reach 36 mph, the Army would not accept it. In order to achieve this goal the new airplane had a smaller wing area, smaller tail areas and was lighter in weight than the aircraft flown in 1908. Also, the machine had a much more powerful engine, which was capable of developing 32 horsepower. The design paid off!

On July 27, following three weeks of testing, Orville Wright with Lt. Frank Lahm as a passenger, flew for a world record

Figure 111

1 hour, 12 minutes while circling the Ft. Meyer drill field 79 times, FIGURE 112 and FIGURE 113. This flight fulfilled the Army's duration specification.

On July 30, Orville was ready to make the official speed flight with Lt. Benjamin Foulois as the passenger/observer. The flight course went south 5 miles, around a cable-suspended balloon on Shooter's Hill in Alexandria, Virginia, and back to Ft. Meyer. The Flyer nearly ready to round the balloon is shown in FIGURE 113. The flight began with the plane flying across the starting line at full speed. Lt. Foulois deducted the time required to turn around the suspended balloon before the return trip. The official speed was the average of both flight segments. The flight covered the distance at an average speed of 42.58 mph. With a base contract price of $25,000 and a two-mile per hour bonus, the Wrights were awarded $30,000 for their efforts. The successful completion of this contract was a fulfillment of the Wrights dream.

## Orville In Germany

In August 1909, Orville and Katharine sailed for Germany en route to Berlin to conduct demonstration flights for the German Wright Company and to teach two of its employees how to fly and how to instruct future customers.

From there they went to Potsdam to provide demonstration and exhibition flights there. Crowds of 200,000 German people, clamoring to see an airplane fly, greeted Orville. It was an exciting time. Germany's Kaiser Wilhelm met with the Wrights and saw them fly. The Kaiser's son, Crown Prince Frederich Wilhelm, became a friend of Orville's and he took the Crown Prince for a 15-minute flight. Afterward, Frederic gave Orville a gold stickpin with the letter "W" formed by diamonds and surrounded with rubies. He said that the "W" could stand for Wright just as it did for Wilhelm. Thus, the Crown Prince Frederich, shown in FIGURE 114, became the first member of a royal family ever to fly in an airplane While in Germany, Orville pushed his airplane to its maximum performance and on one occasion established an altitude record of 1637 feet.

Figure 112 Courtesy of Special Collections and Archives, Wright State University

## Hudson-Fulton Centennial

In late September, Wilbur accepted a contract to fly during the Hudson Fulton Celebration, commemorating the 100th anniversary of Robert Fulton's invention of the steamboat and the 300th anniversary of Henry Hudson's entry into New York harbor. Wilbur received $15,000 for a flight around the Statue of Liberty from Governor's Island. Then on October 4, Wilbur was offered $5,000 more for flying a twenty-mile circuit from Governor's Island, up the Hudson River to Grant's Tomb, and back again over the crowds of ships in the river celebrating the event. More than one million people saw Wilbur's flight that day. For these flights, Wilbur added something

FIGURE 112
Orville and Lt. Frank Lahm in the aircraft prior to the official one-hour duration flight required by the Army contract.

FIGURE 113
Orville and Lt. Benjamin Foulois starting the official ten-mile seed flight, during which Orville achieved an average speed of 42. 58 MPH.

Figure 113 Courtesy of Special Collections and Archives, Wright State University

Die vom "BERLINER LOKAL-ANZEIGER" veranstalteten Flugvorführungen — ORVILLE WRIGHTS in Berlin.

Orville Wright    Kronprinz

Figure 114 Courtesy of Special Collections and Archives, Wright State University

A NEW KIND OF GULL IN NEW YORK HARBOR

Figure 115 Courtesy of Carillon Historical Park Archives

new to his airplane: a bright red canoe, as shown in FIGURE 115 while flying around the Statue of Liberty. For added protection, he had laced a cover over the top of the canoe to prevent the boat from swamping in case he was forced to land in the water.

## College Park, Maryland

The final requirement for fulfilling the Army Contract was to train two soldiers to fly. On October 6, 1909, Wilbur went to College Park, Maryland, to complete contractual obligations. Wilbur moved flight-training operations from Ft. Meyer to College Park, because Ft. Meyer was too small for training. The Army selected Lt. Lamb and Lt. Frederick Humphreys for the training. Lt. Foulois previously had been selected, but duty called him elsewhere. Foulois returned just in time to receive three hours of training from Wilbur. The training went well, and Wilbur made about fifty-five flights during this session. During this time, Wilbur established a 46 mph speed record, his last flight record. He also took Mrs. Ralph Van Deman, a friend of Katharine's, for a ride. She became the first American woman to fly from American soil. These were Wilbur's last public flights and were nearly his last flights as pilot.

## The American Wright Company

Convinced of the need for developing a strong manufacturing entity, on November 22, 1909, Wilbur and Orville Wright sold their patent rights and expertise to the American Wright Company, a newly-formed New York Corporation organized expressly to operate all aspects of the Wright aircraft manufacturing business. The brothers received $100,000 cash outright, one-third of the $1 million worth of the company's capital stock, plus a 10 percent royalty on every machine built and sold. The new firm would be responsible for prosecuting patent infringers and would assume all expenses involved. The company also would solicit and manage all licensing activity. Wilbur was elected president, and a New York financier, Andrew Freedman and Orville, were elected vice presidents. The distinguished board of directors included August Belmont, Robert J. Collier, Cornelius Vanderbilt and Russell Alger.

# Chapter 14 – Epilogue

Beginning in 1909 and continuing from 1910 through 1912, Wilbur Wright's primary activity was defending the Wright patents through lawsuits. No more would his fertile mind be available to add to the scientific knowledge of aeronautics. Originally, the brothers' goal had been to maintain a moderately comfortable standard of living by selling reasonably priced licenses to aircraft builders and then to pursue a life of scientific study and experimentation in aeronautics. But when Glenn Curtiss and many others in America and abroad tried to circumvent the Wright patents or just outright infringed on the Wright patent claims, the Wrights felt morally justified in protecting their rightful property vigorously in the courts. Soon, Wilbur became an expert in explaining the intricate technology that he and his brother had invented in a way that laymen and the courts could understand.

## KEY NUMBER TWENTY-FOUR • TEACH OTHERS TO FLY

Orville, who also at times helped in the court cases, also was very busy. Orville was left to oversee the work in the Wright factories, operate the Wright Flying School and lead the exhibition team established at Huffman Prairie. Teaching others to fly was the twenty-fourth Key to the Wright Brothers' success. Lt. Foulois, Lt. Lamb and Lt. Henry H. "Hap" Arnold were three of the distinguished students of the Wrights. All became Generals in the Army Air Force, and Arnold became a five-star General of the Army Air Force during World War II. Also the first Naval aviators, Lt. Kenneth Whiting and Lt. John Rodgers were instructed at Huffman Prairie. Orville estimated that 115 people learned to fly there. . Teaching others to fly was the twenty-fourth Key to the Wright Brothers' success.

## KEY NUMBER TWENTY-FIVE • MANUFACTURE & SALE OF AIRPLANES

To accommodate the incoming airplane orders, the Wright Company built modern factories having specialized prefabrication areas such as the machine shop, wood shop, and a facility in which wings were covered, as shown in FIGURES 124, 125 and 126 respectively. In 1910, they offered a new Wright Model B airplane having the tail in the rear. FIGURE 127 shows a Model B coming off the assembly line. This was the Wrights' first mass-produced airplane. Manufacturing and selling airplanes was the twenty-fifth Key to the Wright Brothers' success.

### Finale

By 1911, the Wrights had become successful businessmen, and Orville and Katharine decided they needed a new home. They purchased 17 acres of land on a wooded hillside south of Dayton in the suburb of Oakwood. Plans were drawn for the home in early 1912. Wilbur was deeply engrossed in several patent suits at the time and did not show much interest. When the drawings were submitted for his approval, Wilbur indicated he thought there was a lot of wasted entry and hall space but he

FIGURE 116
Wright factory machine shop

Figure 116 Courtesy of Special Collections and Archives, Wright State University

Figure 117 Courtesy of Special Collections and Archives, Wright State University

Figure 118 Courtesy of Special Collections and Archives, Wright State University

Figure 119 Courtesy of Special Collections and Archives, Wright State University

would go along with anything Katharine and Orville wanted. All he wanted was his own bedroom and bathroom. Hawthorn Hill, the Wrights' new home, is shown in FIGURE 128.

But Wilbur Wright never lived in that home. The patent wars were all-consuming and Orville said Wilbur would come home "with his face white" from exhaustion. He returned from a trip to Boston on May 2, 1912 not feeling well. The next day he was feverish and went to bed. Doctors came, and eventually gave the shocking diagnosis - typhoid fever! They had no drug to counteract it in those days. One either fought it off with one's natural resistance or perished. Wilbur's resistance was completely gone. As the month rolled by he became worse and worse until on May 30, 1912, he died. His father's diary carries a beautiful, succinct eulogy:

> "This morning at 3:15, Wilbur passed away, aged 45 years, 1 month, 14 days. A short life, full of consequences. An unfailing intellect, imperturbable temper, great self-reliance and as great modesty, seeing the right clearly, pursuing it steadily, he lived and died."

Figure 120 Courtesy of Carillon Historical Park Archives

It is interesting to note that Wilbur Wright, died exactly 13 years to the day after he wrote his letter to the Smithsonian Institution at the beginning of his aeronautical studies in which he wrote,

> "I wish to avail myself of all that is already known and to add my mite to help on the future worker who will attain final success."

Wilbur and Orville had only 13 years together to make their contribution, and what a contribution they made! Their accomplishments are summarized below:

**Wright Brothers Accomplishments**
- First True Aeronautical Engineers
- First to realize Importance of 3-D Flight Control
- First to Build a Successful 3-D Control System
- First Test Pilots
- First to use Wind Tunnel for:
    Correct Defective Aeronautical Data
    Optimize a Specific Airplane Design
- First to Design an Airplane Propeller
- Designed and Built Aircraft Engine that was beyond the State-of-the-Art
- First to Teach Themselves to Fly a Powered aircraft
- Built Aircraft Factories
- Manufactured Civilian and Military Aircraft
- Taught Others to Fly

FIGURE 117
Wright factory wood shop

FIGURE 118
Wright factory wing covering facility

FIGURE 119
Wright B Flyer at end of assembly line

FIGURE 120
Hawthorn Hill, the Wrights' new home in Oakwood, a suburb south of Dayton

FIGURE 121
Wilbur Wright, born April 16, 1867, died May 30, 1912

Figure 121 Courtesy of Special Collections and Archives, Wright State University

Figure 122

They did it all. Not only did they develop the first successful airplane, they contributed mightily to the development of what has become one of our strongest industries: the aircraft industry.

Wilbur grossly understated his contribution. He added his 'might,' not his "mite" and together with his brother, became that fabled "future worker" who gave the world the airplane.

Orville was devastated by Wilbur's death and gradually his interest in the aviation business faded. However, Orville quickly grasped the baton relinquished by his brother and actively fought the patent battles until they were resolved in the Wright's favor in about 1915. And, once and for all, he finally resolved the controversy with the Smithsonian Institution, which had once fraudulently claimed that Samuel P. Langley's Aerodrome was "capable" of flight before the Wrights actually flew. The Smithsonian finally proclaimed in writing that the Institution had falsely promulgated its earlier statement, and publicly proclaimed that the Wrights were without a doubt the first to fly a manned, heavier-than-air, powered, controllable flying machine; i.e. airplane.

With these controversies out of the way, and after selling his aircraft business, Orville became America's elder statesman of aeronautics and enjoyed his laboratory, family, and vacation home on Lambert Island in Georgian Bay, Canada.

Orville lived to smell the roses. Here in FIGURE 123 he is participating in the dedication of a large stone-marker at the point where the first flights began. He saw the rapid advancement in military airpower during two world wars and witnessed the boom in civilian transportation after World War II. He witnessed the development of both jets and rockets and their subsequent first test flights. All of this he saw in the short 45-year period following his famous flight on that cold day at Kitty Hawk, North Carolina.

Orville was active and alert his entire life. He had his second heart attack while working in his laboratory at 15 North Broadway. His death followed three days later on January 30, 1948. He was in his 77th year when he died.

Figure 123

Figure 124

Today, we can see the original 1903 Kitty Hawk "Wright Flyer" at the National Air and Space Museum, Smithsonian Institution in Washington, D.C. Normally it is displayed as shown in FIGURE 124 suspended as if in flight in its place of honor at the museum entrance in front of the Milestones of Flight Gallery. For the Centennial of Flight Celebration, the Flyer was lowered to floor level where visitors could see all of it clearly along with a special exhibit honoring the Wright Brothers.

FIGURE 122
Orville Wright, Born August 19, 1871, Died January 30, 1948

FIGURE 123
Orville at the dedication of the First Flight Marker at Kitty Hawk

FIGURE 124
1903 Wright Flyer in its place of honor in the Milestones of Flight Gallery, National Air and Space Museum, Smithsonian Institution.

FIGURE 125
Boulder marking the location of the first flights and the distance markers for each of the four flights flown on December 17, 1903 at Kill Devil Hills, North Carolina.

FIGURE 126
Reproductions of the hangar and living quarters used at Kill Devil Hills by the Wright Brothers in 1903.

FIGURE 127
Wright Brothers National Memorial at Kill Devil Hills, North Carolina.

Figure 125

Then, when we visit the Wright Brothers National Memorial at Kill Devil Hills, North Carolina, we can see the huge stone marker placed there in 1928 along with the track and signposts, FIGURE 125, indicating the distance of each of the Wright Brothers' famous four flights in December 1903. Nearby, as in FIGURE 126, we can see modern reproductions of the buildings Wilbur and Orville built there.

But best of all atop the Big Hill at Kill Devil Hills, which has been covered with grass to prevent the hills erosion, is the eye-catching monument, shown in FIGURE 127, that sparkles in the sunlight as its long finger directs our gaze skyward. Approaching closer to its base, we can read the following inscription:

IN COMMEMORATION OF THE CONQUEST OF THE AIR
BY THE BROTHERS WILBUR AND ORVILLE WRIGHT
CONCEIVED BY GENIUS, ACHIEVED BY
DAUNTLESS RESOLUTION AND UNCONQUERABLE FAITH

Figure 126

With Orville on one side and Wilbur on the other looking out towards the sea, it is a magnificent memorial to these immortal men:

Wilbur and Orville Wright

Figure 127

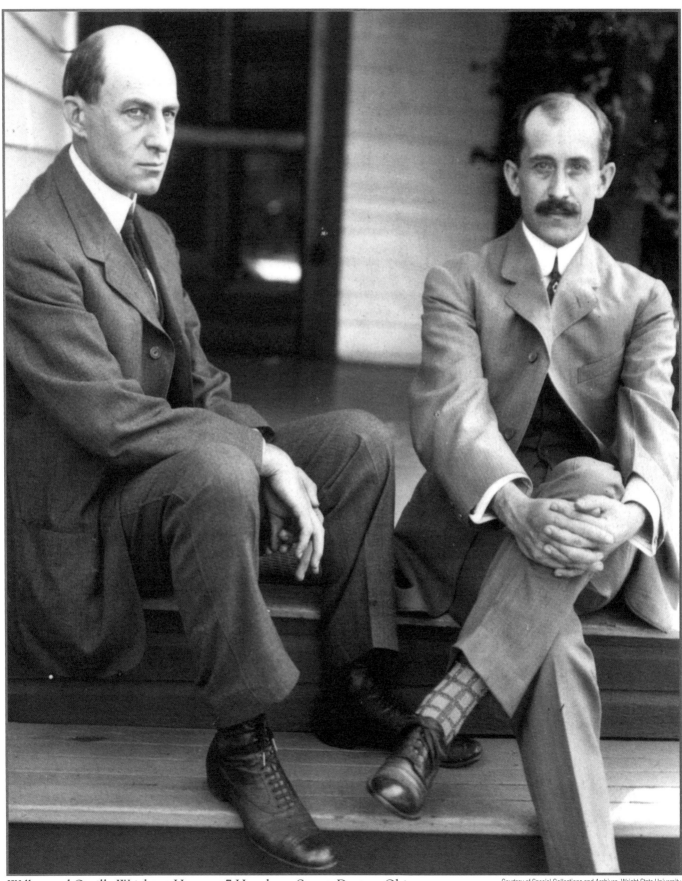

Wilbur and Orville Wright at Home at 7 Hawthorn Street, Dayton, Ohio

Chapter 15
– Tracing Their
Footsteps to Kitty
Hawk

# To celebrate the Centennial of the Dawning of the Age of Flight

My brother and I left Dayton, Ohio, bound for Kitty Hawk, North Carolina to test our glider on the large sand dunes there. No, we aren't the Wright Brothers! It was December 2003, and we are the Whitford brothers, Bob and Russell, the sons of the author of this book. The main purposes of this trip were to follow some of the footsteps of the Wrights as they journeyed to Kitty Hawk and to film the first Flight Centennial Celebration events that are depicted in this chapter. By bringing a radio-controlled glider with us, we also could experience the thrill of flying from the same 'Big Hill' from which the Wrights flew.

## Journey from Dayton to Kill Devil Hills

As we left Dayton, winter's first snow had blanketed the fields and a lone coyote bid us farewell from the roadside. The twisting roads through the West Virginia mountains gave way to the pine country of Virginia, and then we arrived at our first destination: Elizabeth City, North Carolina.

The Wrights had taken three trains and a steamship before arriving at the Norfolk and Southern train station at Elizabeth City. There on Water Street we found Mariners' Wharf, the Wrights' traditional embarkation place for their boat trip across Albemarle Sound to Kitty Hawk. They purchased most of their groceries and supplies near here and had them shipped to Kitty Hawk from this wharf.

After crossing Albemarle Sound by bridge, we found this marker at the former site of the Kitty Hawk Wharf, where the Wrights disembarked. Kill Devil Hill is about 4 miles from here.

Mariners' Wharf at Elizabeth City.

Sign at Mariners' Wharf.

Map of Outer banks area.

Bob and Russell Whitford at the location of the Kitty Hawk Wharf site.

Direction of Kill Devil Hills from Kitty Hawk Wharf.

From the Kitty Hawk Wharf we made our way to the beach and the Kitty Hawk Life Saving Station.

South of the Kill Devil Hills campsite, where the Wright Brothers National Memorial Park is located, one can still see natural, pristine sand dunes at Jockeys' Ridge State Park. This 426-acre park has the largest living sand dune on the East Coast.

Typical beach scene near Kitty Hawk.

Kitty Hawk Life Saving Station, now the Black Pelican restaurant.

Natural sand dunes like those used by the Wrights for their glider tests.

## Tour of the Wright Brothers National Memorial Park

Our final destination was the Wright Brothers National Memorial Park itself, the site of the Wrights campsite and flying area. The Park, established by an act of Congress, includes a 431 acre tract and features a beautiful gleaming white monument on top of the 'Big Hill' from which the Wrights conducted most of their gliding experiments. By planting vegetation on the Big Hill, the War Department stabilized the hill at a height of 80 feet to prevent erosion and shifting of the hill caused by the fierce winds that frequent the area. The 60-foot high monument, dedicated in 1932, is made of North Carolina granite and features a strong beacon light mounted on top that can be seen for miles around. We first climbed the hill to see the monument.

The entire flying field with all of the buildings and facilities erected for the Celebration and the Wrights' flying area can be seen from the top of the Big Hill looking north as well as from the other end of the flying area looking south toward the hill.

While walking around the Park we took a step back in time when we met four visitors, dressed in their authentic 1903 finery, who had come to celebrate the first flight with us.

Continuing our tour of the park we saw various buildings and markers that preserve the history of the Wrights' accomplishments here at Kill Devil Hills.

Photo courtesy of Bob & Virginia Dennison

Pathway to the top of the Big Hill on which the Wright Brothers Monument now stands.

Flying our glider from the Big Hill with Wright Brothers Monument in the background.

Wilbur Wright bust on southeast side of the monument.

Orville Wright bust on the southwest side of the monument.

View of Park from Big Hill looking north.

View of flying area looking south toward the Big Hill.

1903 Visitors. left to right: Jay and Vickie Russon from Greenleaf, Idaho;  Colleen and Boyd Turner from Cary, North Carolina.

Visitor Center which displays two Wright flying machines and documents of their work.

Marker designating the point at which Wilbur Wright made his first attempt to fly the 1903 Flyer on December 14, 1903.

Inside visitor center.

Reproductions of the two sheds the Wrights used for a hangar and living quarters in 1903.

Marker designating the take-off point of Orville Wright's first flight, which traversed distance of 120 feet in 12 seconds.

Marker designating the starting point from which Orville Wright made his first successful flight on December 17, 1903.

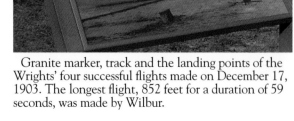

Granite marker, track and the landing points of the Wrights' four successful flights made on December 17, 1903. The longest flight, 852 feet for a duration of 59 seconds, was made by Wilbur.

## Reproductions of Wright Aircraft for Display and for the First Flight Reenactment

Because of their excellent documentation, historians have developed a very good account of what the Wright Brothers did while developing the airplane. However we do not know how they accomplished many of their feats. Ken Hyde, a retired American Airlines captain and an active restorer of historic aircraft, undertook a study during the last few years to learn the secrets behind the Wright Brothers design, development and construction of their airplanes. Forming a non-profit organization known as The Wright Experience, he acquired and studied thousands of documents and pictures concerning the Wrights as well as their existing aircraft, propellers, engines and other artifacts.

Hyde built and flew the Wrights' kite and all of their gliders. When he was finished, he had the information needed to build and fly the most authentic reproduction of the 1903 Wright Flyer that ever has been made. It even has been said that Hyde's 1903 Wright Flyer is more authentic than the original 1903 Wright Flyer in the Smithsonian Institution, because it had to be rebuilt without drawings after it had been severely damaged in 1903. While developing his project, Hyde was careful not to modernize or 'improve' on anything that the Wrights did. His end objective was to produce an aircraft that was as identical to the Wright 1903 Flyer as possible. Sponsored by the Experimental Aircraft Association, the Ford Motor Company and others, his plane was selected to conduct the First Flight Reenactment at Kitty Hawk during the Centennial Celebration.

Wright 1902 glider reproduction in the new Pavilion at the Park.

The 1903 Wright Flyer reproduction in a temporary building. This will replace the Flyer in the Visitor Center.

Hyde also built an exact duplicate of his Flyer under sponsorship by the former president of Gates Lear Jet, Harry Combs, to replace the Wright Flyer currently on display at the Visitors Center here at the Kill Devil Hills Wright National Memorial Park.

Two Wright 1902 gliders and three reproductions of the Wright 1903 Flyer were on display in the Visitor Center, the new Pavilion and one other temporary building. Those shown below were built by Ken Hyde's organization.

## Honoring dignitaries

Fifty-nine descendants from the Wright Family were present at the Celebration. Pictures of some of the Wright Family members from around the country are included. Their progenitors are listed in parentheses.

Various other aviation pioneers were honored. In addition to honoring one hundred aviation pioneers of all time, nineteen living aviation pioneers from every aspect of aviation also were honored. Finally, two people were inducted into the Paul Edward Garber First Flight Shrine.

Paul Edward Garber was the first curator of the National Air and Space Museum, Smithsonian Institution and was responsible for acquiring about half of the famous airplanes that are displayed there during his 73 years of service there. He was a founding trustee of the Shrine, named in his honor, which selects people worldwide who have made outstanding contributions to the field of aviation and space exploration.

The Wright Brothers were the first inductees admitted to the Shrine when it was founded in 1966. The Wright Brothers again were enshrined on this centennial of their first flight. Amanda Wright Lane, the great-grandniece of Wilbur and Orville Wright, announced their enshrinement by unveiling a newly commissioned portrait of them.

Steven Wright Family. (Lorin Wright)

Wilkinson Wright (deceased) -- and wife, Marion Wright. (Lorin Wright)

Amanda Wright Lane Family. (Lorin Wright)

Breen Wright and Milton Wright Jr. (Lorin Wright).

Unveiling portrait of the Wright Brothers, 2003 inductees to the First Flight Shrine.

Margaret Steeper Edwards-Brown (Reuchlin Wright); Marianne Miller Hudec (Lorin Wright); Katharine Wright Chaffee (Reuchlin Wright).

Paul Edward Garber, founding trustee of The First Flight Shrine.

Ken Hyde and the 1903 Flyer.

## Preparation for the Centennial Flight

A dedicated team of people led by Ken Hyde worked in a hangar located next to the flying field for several months preparing for the Celebration.

Hyde's group gave special attention to pilot selection and training. Flight simulators were designed, based on wind tunnel tests of the full-scale aircraft, and four pilots were trained. Scott Crossfield, the famous NASA X-15 test pilot, was engaged to supervise the training and final pilot selection. Training included flying the 1902 glider, the Wright Flyer flight simulator and eventually the 1903 Flyer itself. The two pilots finally selected for the flight were Kevin Kochersberger and Terri Queijo.

The Wright Experience crew with the Flyer.

All photos courtesy of Paul Glenshaw of the Wright Experience

Scott Crossfield prepares for flight training in the Wright 1902 glider.

Scott Crossfield (2nd from left) with pilot candidates Terry Queijo (L) Ken Hyde and Kevin Kochersberger (R).

Dr. Kevin Kochersberger is an associate professor of mechanical engineering at the Rochester Institute of Technology who began his flying career by hang gliding from the dunes of North Carolina. He has been an active glider, sailplane and airplane pilot since age 15. He has worked on the construction of the 1903 Flyer and supervised the wind tunnel testing.

Terry Queijo began flying without a machine; i.e. by sky diving and progressing to serious skydiving competition. Eventually she obtained a pilot's license and became a captain on American Airlines 757 and 767 jet transport airplanes flying out of Washington, D.C.

Flyer in hangar with Ken Hyde talking to reporters.

Chris Johnson (pilot candidate), Scott Crossfield and Kevin Kochersberger in discussion during flight training.

Flyer during roll out with spectators surrounding it.

Flyer nearing its destination for test.

Scott Crossfield adjusting front end on rail.

Flyer being rolled on track.

## December 17, Centennial of First Flight Reenactment

The day dawned cloudy, and as predicted a steadily increasing rainstorm began at about 8:00 a.m. and continued for most of the morning. An estimated throng of 45,000 people was present wearing rain gear. Not to be deterred by the rain, President George Bush's entourage of six military helicopters provided an air show in itself. The President stood unprotected from the rain, and said these words to the to the cheering throng:

"Powered flight has advanced in ways that could not have been imagined on December 17, 1903. - A great American journey that began at Kitty Hawk continues in ways unimaginable to the Wright Brothers. One small piece of their Flyer traveled far beyond this field - on Apollo 11- all the way to the Sea of Tranquility on the moon. - Yet always, for as long as there is human flight, we will honor the achievement of a cold morning on the Outer Banks of North Carolina by two young brothers named Orville and Wilbur Wright. - Everyone who was here at that hour sensed that a great line had been crossed and the world might never be the same. -We take special pride in their qualities of discipline, persistence, optimism and imagination. This day, however, is one for recalling an heroic event in the history of our nation and in the story of mankind. Here at the Wright Brothers National Memorial, we remember one small machine and we honor the giants who flew it."

Spectators on Hill waiting for the flight.

Spectators on the other side of the field.

Flyer on track ready for take off.

Flyer after a long run down the track with the front end lifted from the track in flying position.

Flyer at end of attempt being recovered by the ground crew.

Keith Yeorg (Wright descendent), Terry Queijo, Pamela Hyde, Ken Hyde, Amanda Wright Lane, and Steven Wright after flight attempt on December 17, 2003.

A steady rain was falling at the scheduled time for the Centennial reenactment, 10:35 a.m. So the flight was postponed until 2:00 p.m. because the Flyer's weight would increase to a prohibitive level if its wings and tail became rain-soaked.

At about noon, the rain stopped, a window of opportunity opened and the decision was made to fly! The plane was brought out rapidly and placed on the track. Kevin Kochersberger was selected as the pilot for this first try. Both Kochersbberger and Terry Queijo were dressed in clothing similar to that worn by Orville and Wilbur 100 years earlier. The humid weather made starting the engine difficult, but soon it was running. The track was pointing to the south but the wind was variable from south to south-southwest and partly blanketed by the Big Hill. Since the wind was blowing only about 12 mph compared to the 22 to 27 mph gale into which the Wrights launched their Flyer, a 210-foot long track was laid to provide the Flyer more time to accelerate to flying speed.

Unfortunately, the plane did not acquire adequate flying speed. It abruptly dropped back down on the track and the flight ended with the right wing tip splashing into a pool of water. The plane was examined and then wheeled back into the hangar. Two cross braces between the landings skids had broken when the front end of the Flyer smashed down on the track. The engine also was dismantled to clean the breaker points in the cylinders.

Why didn't the airplane fly? There were two reasons. First the engine at the start of the run was turning at only 1000 rpm instead of 1050 rpm. Further, part way down the track one of the cylinders ceased to operate, so the engine was firing on only three of the four cylinders. Thus the engine probably was providing less than 8 horsepower instead of the expected 12 horsepower, which was not enough power to accelerate the Flyer to flying speed. Secondly, Mother Nature did not cooperate. Instead of relatively dry 34° F in which the Wrights flew, Hyde's crew in 2003 faced humid, 60° F air with the barometric pressure probably lower than it was in 1903. This difference in weather conditions could have reduced the 2003 air density and hence the wing's lift and the propeller thrust by as much as 10 percent. Thus the loss in propulsive power caused by a malfunctioning engine and a less-productive propeller coupled with the wing's loss of lift did not provide enough force to lift the plane from the ground.

Replacement of the cross brace took only about 30 minutes. Overhauling the engine to clean the breaker points and dry it out took a little longer. The pictures below show the plane in the hangar while engine overhaul is underway.

Later in the afternoon, a weather front passed through and the wind shifted to the north, eliminating the partial blanketing of the wind by the Big Hill that occurred earlier. At about 3:30 p.m. the repaired plane was brought out for a second flight attempt.

The mechanics cranked the propellers and after several attempts the engine started. This time the engine was running perfectly. However, by the time all was ready, the wind speed had dropped to less than three mph, not enough to assure flight. So the order was given to cut the engine and discontinue efforts to fly. No further attempts to fly the airplane will be made. It will be sent to Greenfield Village in Dearborn, Michigan for display with the Wright home and bicycle shop that were moved to Dearborn from Dayton, Ohio, in 1936.

Greg Cone overhauling the engine in the hangar.

Pilot Kevin Kochersberger talking to spectators watching engine repairs being made.

The Flyer on the track ready for a second attempt.

The Flyer with engine running and pilots in position.

Sequence of images from December 3, 2003 test flight. Note the lift off then the right wing brushing the sand and the flyer lifting off again.

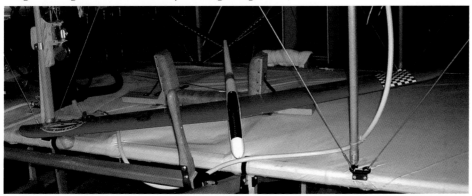

Model Glider lying on the wing of the Wright Flyer.

Photo courtesy of Dave Domingue of the Wright Experience

## The Wright Flyer built by The Wright Experience actually did fly!

The following picture was taken during flight tests of Hyde's 1903 Flyer in early December of 2003. The plane flew, so the quality of Hyde's airplane is not in question. Ken Hyde needed only to have Mother Nature's support to completely fulfill the expectations of the First Flight Centennial Celebration.

Everyone affiliated with the Celebration was aware of the weather risk. With the severe weather conditions usually experienced at Kill Devil Hills during the month of December, it appeared that such a flight would be possible. Unfortunately, the weather was too warm, humid and balmy.

Nevertheless, the Celebration was an excellent festival and the people of North Carolina are to be commended for their outstanding effort in celebrating this world-class event. Just seeing the Flyer on the launching rail and hearing the engine run was a thrill not to be forgotten.

Before we left Kill Devil Hills, we visited Ken Hyde's Wright Flyer once more, and watched while our radio-controlled glider was gently laid upon the Flyer's lower wing. Immediately we realized that the resulting picture was particularly symbolic of mankind's legacy from Wilbur and Orville Wright: the Wright Flyer not only was physically supporting the model glider, but more importantly, the Flyer has provided the supporting technological foundation upon which all subsequent airplanes, including the model glider, have been, are and will be based.

My brother, my father and I relived our trip as we returned to Dayton from Kitty Hawk. We had walked in the footsteps of the Wright Brothers, we had flown a glider from the same hills from which they once flew, and we had paid homage to their achievements. Yet even in this day of aviation advancement, inconceivable to the Wrights in 1903, we are humbled by the fact that no one since 1903 has flown an authentic 1903 Wright Flyer further or longer than Wilbur Wright flew it on December 17, 1903.

The monument on the Big Hill is a perpetual testimony to Wilbur and Orville Wrights' genius, dedication and skill. As their work led the way to the development of the aviation industry that we enjoy today, so does the beacon atop the magnificent white shaft of granite provide perpetual guidance to those who aspire to lead in the future.

The Wright Brothers' Monument at Kill Devil Hills, North Carolina.